SpringerBriefs in Electrical and Computer Engineering

Computational Electromagnetics

AF758335

Series editors

K.J. Vinoy, Bangalore, India
Rakesh Mohan Jha (Late), Bangalore, India

More information about this series at http://www.springer.com/series/13885

Raveendranath U. Nair · Maumita Dutta
Mohammed Yazeen P.S. · K.S. Venu

EM Material Characterization Techniques for Metamaterials

Springer

Raveendranath U. Nair
Centre for Electromagnetics (CEM)
CSIR-National Aerospace Laboratories (CSIR-NAL)
Bengaluru, Karnataka
India

Mohammed Yazeen P.S.
Centre for Electromagnetics (CEM)
CSIR-National Aerospace Laboratories (CSIR-NAL)
Bengaluru, Karnataka
India

Maumita Dutta
Centre for Electromagnetics (CEM)
CSIR-National Aerospace Laboratories (CSIR-NAL)
Bengaluru, Karnataka
India

K.S. Venu
Centre for Electromagnetics (CEM)
CSIR-National Aerospace Laboratories (CSIR-NAL)
Bengaluru, Karnataka
India

ISSN 2191-8112 ISSN 2191-8120 (electronic)
SpringerBriefs in Electrical and Computer Engineering
ISSN 2365-6239 ISSN 2365-6247 (electronic)
SpringerBriefs in Computational Electromagnetics
ISBN 978-981-10-6516-3 ISBN 978-981-10-6517-0 (eBook)
https://doi.org/10.1007/978-981-10-6517-0

Library of Congress Control Number: 2017952001

Printed on acid-free paper

This Springer imprint is published by Springer Nature
The registered company is Springer Nature Singapore Pte Ltd.
The registered company address is: 152 Beach Road, #21-01/04 Gateway East, Singapore 189721, Singapore

To
Late Dr. Rakesh Mohan Jha
Founder Scientist
Centre for Electromagnetics
CSIR-NAL, Bangalore

Preface

Rapid growth in technology demands application-specific materials. Accurate EM characterization of a material is a requisite for the categorization of the materials. Material parameters such as permittivity, permeability, reflection coefficient, and transmission coefficient are the key features in typifying a material. Innovations are happening in the field of material science to be in par with the fast advancing technological developments. One of the noble progress in this field is evolution of metamaterial which is an artificially fabricated material. The unique attributes of metamaterial are its negative intrinsic properties (permittivity and permeability) which are obtained from the different orientation of the inclusions on the substrate. At higher frequencies, the incorporation of metamaterials helps in device size reduction with increased efficiency.

Metamaterials need to be characterized for deploying in various applications. The different characterization techniques for metamaterials are explained in this brief and are organized as follows: Section 1 deals with the introduction to material parameters as well as metamaterial and its importance. Section 2 explains the basics of material characterization and a brief explanation on resonant and non-resonant methods of material characterization. Section 3 details the various techniques of EM material characterization of metamaterials. Finally, Section 4 summarizes the entire brief.

Bengaluru, India

Raveendranath U. Nair
Maumita Dutta
Mohammed Yazeen P.S.
K.S. Venu

Acknowledgements

We would like to thank Mr. Shyam Chetty, Director, CSIR-National Aerospace Laboratories, Bangalore, for giving permission to write this SpringerBrief.

Further, we record our thanks to colleagues Dr. Hema Singh, Dr. Shiv Narayan and, Dr. Balamati Choudhury, for their suggestions and cooperations. We express our sincere thanks to Project Staff of Centre for Electromagnetics, Ms. Suprava M., Ms. Karthika S. Nair, Ms. Vinisha C.V., and Ms. Mahima P., for their consistent support during the preparation of the technical contents of this book. It would have not been possible to bring out this book within a short span of time without consistent support and suggestions of the team at Springer, particularly Swati Meherishi, Kamiya Khattar, and Aparajita Singh.

About the Book

Since metamaterials are typically inhomogeneous and anisotropic, the experimental techniques for EM material characterization of metamaterial structures need to tackle several challenges. Further, the modes supported by metamaterial structures are extremely sensitive to external perturbations. Hence, the measurement fixtures for EM material characterization have to be modified to account for such effect. Considering the above-mentioned aspects, a review of EM material characterization techniques based on waveguide systems, striplines, free space systems, etc., are presented. The salient features of each method are discussed in detail in this book.

Contents

About the Authors

Dr. Raveendranath U. Nair is currently Senior Principal Scientist and Head, Centre for Electromagnetics (CEM), CSIR-National Aerospace Laboratories (CSIR-NAL), Bangalore, India. Dr R.U. Nair has authored/co-authored over 150 research publications including peer reviewed journal papers, symposium papers, and technical reports. He has co-authored a chapter in a book Sensors Update published by Wiley-VCH, Germany, in 2000. The electromagnetic (EM) material characterization techniques developed for his doctoral work were included in the section *Perturbation Theory* in RF and Microwave Encyclopaedia (Vol. 4) published by John-Wiley & Sons, USA, in 2005.

Dr. R.U. Nair received the CSIR-NAL Excellence in Research Award (2007–2008) for his contributions to the EM design of variable thickness airborne radomes. His research interests include electromagnetic design and performance analysis of radomes, radar cross section, frequency selective surfaces (for radomes and RAS), EM material characterization, complex media electromagnetics, and microwave measurements. He took key responsibilities in setting up of three major national facilities (Adaptive Antenna Facility, FSS Characterization Facility, and Electromagnetic Materials Applications Facility) at CEM, CSIR-NAL, Bangalore. He is a life member of Aeronautical Society of India (AeSI), member of ISAMPE, and member of IEEE. Dr. R.U. Nair is also a Professor of the Academy of Scientific and Innovative Research (AcSIR), New Delhi.

Ms. Maumita Dutta obtained M.Tech. in RF and Microwave Engineering from R. V. College of Engineering, Visvesvaraya Technological University, Bangalore. She obtained her B.Tech. (ECE) degree in 2012 from ICFAI University, Tripura. She is a Project Scientist at Centre for Electromagnetics (CEM) of CSIR-National Aerospace Laboratories, Bangalore, and her working area includes metamaterials and material characterization.

Mohammed Yazeen P.S. is currently working as a project scientist with the Centre for Electromagnetics (CEM) of CSIR-National Aerospace Laboratories (CSIR-NAL), Bangalore, India, since February 2015. He was born in 1990 at

Kerala, India. Mohammed Yazeen P.S. obtained his M.Tech. degree in Electronics with specialization in Wireless Technology from Cochin University of Science and Technology (CUSAT), Cochin, India, in 2014. His research interests are design and analysis of streamlined airborne radomes, wireless communication, structurally integrated radiating structures, and EM material characterization.

Mr. K.S. Venu is currently Senior Technical Officer, Centre for Electromagnetics (CEM), CSIR-National Aerospace Laboratories (CSIR-NAL), Bangalore, India. He received the B.E. (Electronics & Communication) Bangalore University, Karnataka, India, in 2001.

Mr. K.S. Venu has authored/co-authored over 30 research publications including journal papers, and technical reports.

His research interests include electromagnetic design and performance analysis of radomes, frequency selective surfaces (for radomes and RAS), EM material characterization, complex media electromagnetics, and microwave measurements. He took key responsibilities in setting up of three major national facilities (Adaptive Antenna Facility, FSS Characterization Facility, and Electromagnetic Materials Applications Facility) at CEM, CSIR-NAL, Bangalore. He is a life member of ISAMPE.

Abbreviations

DNG	Double negative materials
EM	Electromagnetic
ENG	Epsilon negative materials
FSM	Free space measurements
IR	Infrared
MUT	Material under test
MW	Microwave
RAS	Radar absorbing structure
SNG	Single negative materials
SRR	Split ring resonator
TEM	Transverse electromagnetic
VNA	Vector network analyzer

Symbols

ε_r^*	Complex relative permittivity of the test sample
Γ	Propagation constant
λ_0	Free space wavelength
u_d	Relative permittivity of the known dielectric
u_{lr}	Relative permeability in the longitudinal directions
μ_0	Absolute permeability
μ_r^*	Complex permeability of the test sample
μ_t	Unambiguous complex permeability
μ_{tr}	Relative permeability in the transversal directions
μ_x	Permittivity in the x direction
μ_y	Permeability in the y direction
a	Board dimension of the waveguide
c	Speed of light
d	Thickness
f	Frequency of the signal
f_{co}	Cutoff frequencies of an empty waveguide
f_c	Cutoff frequencies of an empty waveguide filled with material
h_a	Height
h_s	Height of the sample
K_d	Propagation constant in the known dielectric
K_m	Propagation constant in the unknown metamaterial sample
W	Width
Y_1	Shunt admittance
Y_2	Shunt admittance
Z_d	Characteristic impedance in the known dielectric
Z_m	Characteristic impedance in the unknown metamaterial sample
Z_0	Characteristic impedance of the cell
Z_w	Wave impedance

List of Figures

List of Tables

Chapter 1
Introduction

An interaction between the electromagnetic fields and materials occurs in all RF and microwave applications, be it solid, liquid or gas. Electromagnetic waves always penetrate through materials to some extent, even if it a good conductor or superconductor. Consequently, all materials can be called as dielectrics which in other words mean that electric field can pass through them. Knowing the propagation of electric and magnetic field through the materials as well as the change in behaviour of waves at interfaces are requisites for the RF designing of materials. In RF and microwave, parameters capturing the macroscopic significances of all the microscopic prodigies of a material are taken into consideration. Among these parameters, the most important ones are complex electric permittivity, ε^* and complex magnetic permeability μ^*.

Absolute permittivity can be obtained by multiplying Relative permittivity with permittivity of free space, $\varepsilon_0 = 8.8542 \times 10^{-12}$ Fm^{-1}. The SI unit for absolute permittivity is farads per metre whereas relative permittivity is dimensionless. Relative permittivity has two components, i.e. a real component and an imaginary component and can be represented as

$$\varepsilon^* = \varepsilon' - j\varepsilon''$$

where, $j = \sqrt{-1}$, ε' represents the real part of relative permittivity and ε'' represents its imaginary part. Suppose a dielectric material is placed between plane parallel electrodes so as to form a capacitor, then ε' depicts the capacitive part of admittance, Y, and ε'' typifies the conductive part or in other words the lossy part of admittance, Y. The values of ε' and ε'' depend on several parameters like temperature, relative humidity and frequency.

Analogous to relative permittivity ε^*, its corresponding magnetic parameter is complex relative magnetic permeability, μ^* which can also expressed in terms of real and imaginary components.

R.U. Nair et al., *EM Material Characterization Techniques for Metamaterials*,
SpringerBriefs in Computational Electromagnetics
https://doi.org/10.1007/978-981-10-6517-0_1

$$\mu^* = \mu' - j\mu''$$

Metamaterials are artificially fabricated structures with unique material properties (like negative permittivity, negative permeability, etc.), that are seldom found in naturally occurring materials. As metamaterials are structures of periodically arranged unit cell, it has close resemblance with natural crystal. An important feature of metamaterials is that it is possible to tune their electric permittivity and magnetic permeability for specific applications. Metamaterials usually gain their properties from structure rather than composition, the small inhomogeneity is used to create effective macroscopic behaviour.

The research and development that has already taken place in the area of metamaterials, has offered promising components and subsystems that could potentially surpass the limitations of current technology. The immense technological potential of metamaterials has drawn attention of many researchers working in microwave (MW), millimetre wave (mmw) and infrared (IR) frequency regimes for developing novel components and subsystems for aerospace applications. Metamaterials are commonly known as double negative materials (DNG) because the electric permittivity (ε) and magnetic permeability (μ) are simultaneously negative for these materials.

Metamaterials have wide range of applications. An invisibility cloak can be made using metamaterial by effectively bending the light. Metamaterials can also help in ultra-fast processing of data. As the size of chip is smaller, it can be used for establishing broadband connection in trains, planes, cars, etc. Smartphone lenses are also made up of metamaterials. Some applications of metamaterials related to aircraft are array feed networks, phased array antennas, antenna radomes, struts in reflector antennas, antenna ground planes, etc.

The EM design and development of metamaterial based hardware components/ subsystems (like antennas, radomes, radar absorbing structures (RAS), etc.) depends extensively on the intrinsic properties of metamaterials. The reliability of such designs depends on the accuracy to which the EM material parameters of the metamaterials have been measured and hence there is a growing need for the development of novel metamaterial characterization techniques. A review of measurement methods for EM material characterization of metamaterials based on open domain sources has been presented in this document to assess the efficacy of each method.

Chapter 2
Basics of Material Characterization

In this section, the basic principle for material characterization, which is finding or measurement of permittivity and permeability of a material is discussed.

According to microwave theory, material characterization can be categorized into non-resonant methods and resonant methods. If the electromagnetic properties of a material over a frequency range is to be known then non-resonant methods are used, whereas to know the dielectric properties of a material for a specific frequency, resonant methods are used. To know the exact properties of materials both resonant and non-resonant methods are used hand in hand. The properties of materials over a frequency range is improved using the non-resonant methods with the help of the information received by using resonant methods for the same material at a particular frequency.

2.1 Non-resonant Methods

In this method, the properties of the materials are inferred from their impedance and wave velocity. The propagation of electromagnetic waves from free space to samples results in changes in characteristic impedance and wave velocity of the material. Due to these changes, partial reflection of electromagnetic wave occurs, at the interface between the two materials.

Non-resonant method can be further categorized into two methods, they being reflection method and transmission/reflection method. In reflection method, estimation of material properties are done on the basis of reflection from the sample. In transmission/reflection method, estimation is done on the basis of reflection from the sample and transmission through the sample, in this method, the electromagnetic energy needs to be guided towards the material and then acquiring the reflected energy from the material as well as through the material.

R.U. Nair et al., *EM Material Characterization Techniques for Metamaterials*,
SpringerBriefs in Computational Electromagnetics
https://doi.org/10.1007/978-981-10-6517-0_2

2.2 Resonant Method

The accuracy and sensitivity using resonant methods are higher as compared to non-resonant methods and are most likely for low-loss sample. The two categories of resonant methods are resonator method and resonant-perturbation method. In the former method, the permittivity and permeability helps in the calculation of resonant frequency and quality factor of dielectric resonator for the given dimensions. The later method is rooted from resonant-perturbation theory. According to this theory when changes are made to electromagnetic boundaries of a resonator, a visible variation is observed in the resonant frequency and quality factor of the sample. Analysing these changes the properties of the sample can be deduced.

Chapter 3
EM Material Characterization Techniques

It is technology that fixes the use of electromagnetic materials, but again it is science that infers the behaviour of the material. Material's response towards electromagnetic fields is decided by the displacement of material's free and bonded electrons by the electric fields in addition to the atomic moment's orientation by magnetic fields.

It is very important to know about the wide-ranging properties of electromagnetic materials as it can help in validating the measured outcome and also the possible reasons for the inaccuracy that one can come across during material characterization. The various electromagnetic materials include dielectric materials, artificial materials, semiconductors, magnetic materials and conductors.

In the case of metamaterials, the entire characterization of effective medium can be achieved by recovering all the components of the permeability and permittivity tensors and accounting for both spatial and time dispersion. The competence of the method in characterizing the anisotropy is particularly useful in the design and realization of metamaterial-based structures, which facilitates control of all the components of constitutive parameters. In view of this, a brief review of EM material characterization techniques based on waveguide systems, striplines and-freespace systems are presented in the subsequent sections.

3.1 Waveguide-Based Methods

In this method, rectangular waveguides are used to measure the permittivity and permeability of metamaterials. The characterization of metamaterials involves the definition of a unit cell, which in turn defines the structure of metamaterial. Here, different types of unit cells corresponding each artificial structures are defined and simulated to compute the reflection and transmission characteristics of the structure for the determination of the dielectric and magnetic parameters of the metamaterial. This method mainly focuses on effective dielectric and magnetic properties of the

R.U. Nair et al., *EM Material Characterization Techniques for Metamaterials*,
SpringerBriefs in Computational Electromagnetics
https://doi.org/10.1007/978-981-10-6517-0_3

metamaterial when it is oriented perpendicular to the direction of propagation within the waveguide system (Damascos et al. 1984). The measurement setup includes a rectangular waveguide in which a sample of metamaterial is fitted into the waveguide cross section. The reflection and transmission coefficients are measured using the network analyzer over the specified frequency range corresponding to the waveguide system.

3.1.1 Waveguide-Based Transmission/Reflection Measurement Technique

In order to measure the complex permittivity and permeability of metamaterials, Transmission/Reflection (T/R) measurements are carried out by means of rectangular waveguide system (Buell and Sarabandi 2002). In this method, a sample metamaterial block with cross-sectional dimensions of rectangular waveguide is positioned in the waveguide system as shown in Fig. 3.1.

With the help of vector analyzer, the measurements for the complex reflection coefficient (Γ) and transmission coefficient (T) at the required frequency/frequency ranges are carried out. It is noted that the cross-sectional dimensions of the sample must match to that of the waveguide dimensions. Proper choice of thickness depends on a number of factors. However, it is advisable to keep the sample thickness less than a wavelength in order to avoid phase ambiguity.

If we assume that a wave is incident in Region I of Fig. 3.1, the field distributions in all the three regions (i.e. Region I, Region II and Region III) are given by the following equations:

$$E_1 = \exp(-\eta_o z) + \Gamma \exp(\eta_o z) \tag{3.1}$$

$$E_2 = C_1 \exp(-\eta_g z) + C_2 \exp(\eta_g z) \tag{3.2}$$

$$E_3 = T \exp(-\eta_o z) \tag{3.3}$$

Taking into account the boundary conditions of the material (front and back surfaces), the unknown coefficients (C_1 and C_2) can be analytically determined.

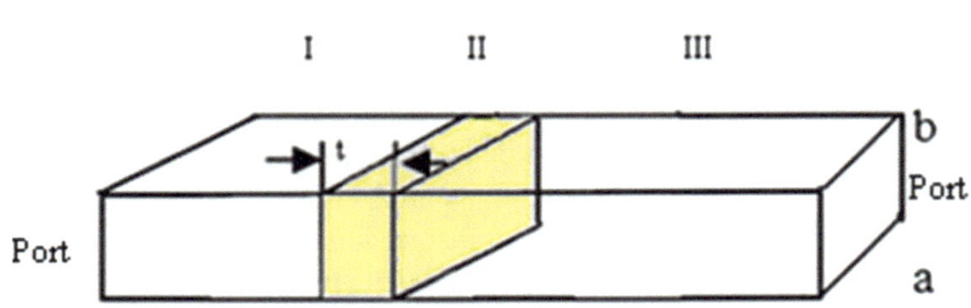

Fig. 3.1 Schematic of waveguide system loaded with material under test

The reflection and transmission coefficients of the metamaterial can be determined by the following equation:

$$\lambda_g^m = \frac{2a\lambda_o}{\sqrt{4\mu_r\varepsilon_r a^2 - \lambda^2}} \tag{3.4}$$

$$\lambda_g = \frac{2a\lambda}{\sqrt{4a^2 - \lambda^2}} \tag{3.5}$$

$$\Gamma = \frac{i\sin\left(2\pi t/\lambda_g^m\right)\left(\frac{1}{\lambda_g^{m2}} + \frac{2\mu^2}{\lambda_g^2}\right)}{i\left(\frac{1}{\lambda_g^{m2}} + \frac{\mu^2}{\lambda_g^2}\right)\sin\left(2\pi t/\lambda_g^m\right) - \frac{g\mu\cos\left(2\pi t/\lambda_g^m\right)}{\lambda_g^m\lambda_g}} \tag{3.6}$$

$$T = \frac{\mu}{\left[i\exp\left(\frac{i2\pi}{\lambda_g^m}\right)\left(\frac{\lambda_g}{\lambda_g^m}\right)^2\sin\left(\frac{2\pi t}{\lambda_g^m}\right) + i\exp\left(\frac{i2\pi}{\lambda_g}\right)\left(\frac{\lambda_g^m}{\lambda_g}\right)^2\mu^2\sin\left(\frac{2\pi t}{\lambda_g^m}\right) - 2\exp\left(\frac{i2\pi}{\lambda_g}\right)\mu\cos\left(\frac{2\pi t}{\lambda_g^m}\right)\right]} \tag{3.7}$$

By separating the complex reflection and transmission coefficients into corresponding real and imaginary terms, one can obtain the values of ε', ε'', μ' and μ'' of metamaterial sample.

3.1.2 Waveguide-Based Retrieval Method

In another reported method (Chen et al. 2006), different sets of independent scattering data corresponding to different orientations of the metamaterial sample are measured in the frequency range for the dominant TE_{10} mode of rectangular waveguide. The whole procedure is first verified by a slab of homogenous frequency dispersive material with known material parameters and then verified by an anisotropic metamaterial composed of SRR structures.

The experimental setup consists of two coaxial-to-waveguide adapters connected to a rectangular waveguide loaded with the slab of the material under test, as shown in Fig. 3.2. The metamaterial sample having thickness '*d*' completely fills the cross section of the waveguide. It is noted that the distances between the coaxial-to-waveguide adapters to the sample are large enough so that the higher-order evanescent modes are considerably attenuated before reaching the sample under test. When the reference plane coincides with the first face of the slab, S_{11} is equal to the reflection coefficient. Then, S_{21} is related to the transmission coefficient T by the relation

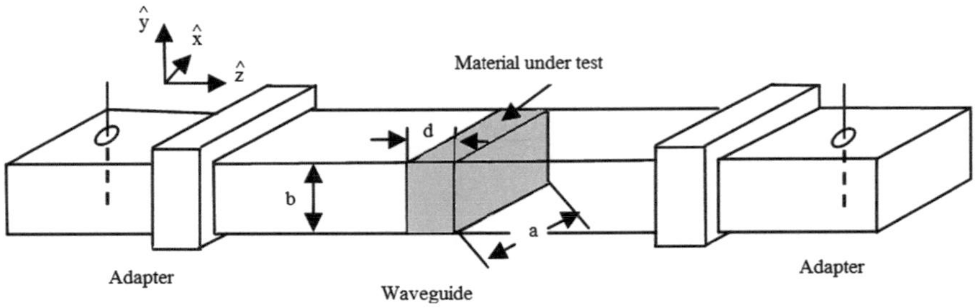

Fig. 3.2 Rectangular waveguide systems for EM material characterization of metamaterial slab

$$S_{21} = T\exp(ik_{oz}d), \tag{3.8}$$

where k_{oz} is the longitudinal wave number of the incident wave.

During measurement using vector network analyzer, the two-port calibration is initially performed at the end faces of waveguide system. Then the calibration at the two reference planes corresponding to the two interfaces between the sample and the air is carried out. The permittivity and permeability tensors of the material can be retrieved using the measured *S*-parameters.

The constitutive relations of the material are given by the following equation:

$$\overline{D} = \overline{\overline{\varepsilon}}\overline{E}, \quad \overline{B} = \overline{\overline{\mu}}\overline{H} \tag{3.9}$$

where the parameter tensors have the following forms in the principal system (e_1, e_2, e_3): $\overline{\overline{\varepsilon}} = \varepsilon_0[\varepsilon_1\varepsilon_2\varepsilon_3]$, $\overline{\overline{\mu}} = \mu_0\text{diag}[\mu_1\mu_2\mu_3]$. The principal axes are the three coordinate axes e_1, e_2, and e_3. In order to get all the constitutive parameters, different samples of the material under test are required for the measurement. Here, the parameters under consideration are μ_1, μ_2 and ε_3. Correspondingly, two independent measurements (M1 and M2) are necessary to estimate these three parameters.

M1: The axes e_1, e_2 and e_3 of the sample slab are along the direction of $\hat{x}$, $-\hat{z}$ $\hat{y}$, respectively.

M2: The axes e_1, e_2 and e_3 of the sample slab are along the direction of $\hat{z}, \hat{x}, \hat{y}$, respectively.

The refractive index n and the impedance Z can be evaluated from the measured *S*-parameters given below:

$$Z = \pm\sqrt{\left\{\left[(1+S_{11})^2 - S_{21}^2\right]/\left[(1-S_{11})^2 - S_{21}^2\right]\right\}} \tag{3.10}$$

$$\exp(ink_{oz}d) = X \pm i\sqrt{1-X} \tag{3.11}$$

where $X = \left(1 - S_{11}^2 + S_{21}^2\right)/2S_{21}$.

Here, n and Z are

$$n_a = \sqrt{k_o^2 \varepsilon_3 \mu^2 - k_x^2 \mu_1} \Big/ \sqrt{k_o^2 - k_x^2} \tag{3.12}$$

$$Z_a = \mu_1 \sqrt{k_o^2 - k_x^2} \Big/ \sqrt{k_o^2 \varepsilon_3 \mu_2 - k_x^2 \mu_1 / \mu_2} \tag{3.13}$$

$$n_b = \sqrt{k_o^2 \varepsilon_3 \mu^2 - k_x^2 \mu_2 / \mu_1} \Big/ \sqrt{k_o^2 - k_x^2} \tag{3.14}$$

$$Z_b = \mu_2 \sqrt{k_o^2 - k_x^2} \Big/ \sqrt{k_o^2 \varepsilon_3 \mu_2 - k_x^2 \mu_2 / \mu_1} \tag{3.15}$$

where $k_x = \pi/a$ represents the transverse wave number in the rectangular waveguide system.

The subscripts 'a' and 'b' represent the results that are calculated from the measurements of M1 and M2, respectively.

Using Eqs. (3.12)–(3.15), one gets

$$\mu_1 = n_a z_a, \quad \mu_2 = n_b z_b \tag{3.16}$$

ε_3 can be calculated from Eqs. (3.12) and (3.16), or from Eqs. (3.14) and (3.16), respectively,

$$\varepsilon_{3a} = \left(n_a^2 k_{oz}^2 + k_x^2 \mu_1\right) / k_o^2 \mu_1 \tag{3.17}$$

$$\varepsilon_{3b} = \left(n_b^2 k_{oz}^2 + k_x^2 \mu_2 / \mu_1\right) / k_o^2 \mu_2 \tag{3.18}$$

Likewise, the other three constitutive parameters μ_3, ε_1 and ε_2 can be evaluated based on two additional measurements (M3 and M4).

M3: The axes e_1, e_2 and e_3 of the material are along with the coordinates $(-\hat{z}, \hat{y}, \hat{x})$. This measurement is required for the estimation of μ_3 and ε_2.

M4: The axes e_1, e_2 and e_3 of the material are along with the coordinates $(\hat{y}, \hat{z}, \hat{x})$. This measurement is required for the estimation of ε_1.

This waveguide method is applicable to various metamaterial structures. It is noted that the measurement procedure is easy to carryout and no stringent requirements are needed for the sample under test.

3.1.3 Waveguide Verification Design Procedure

Calibration (Lozano-Guerrero et al. 2010; Williams et al. 2003) of a system is important and is the first step for any measurement systems. Calibration helps in attaining confidence over the system by providing known results under suitable

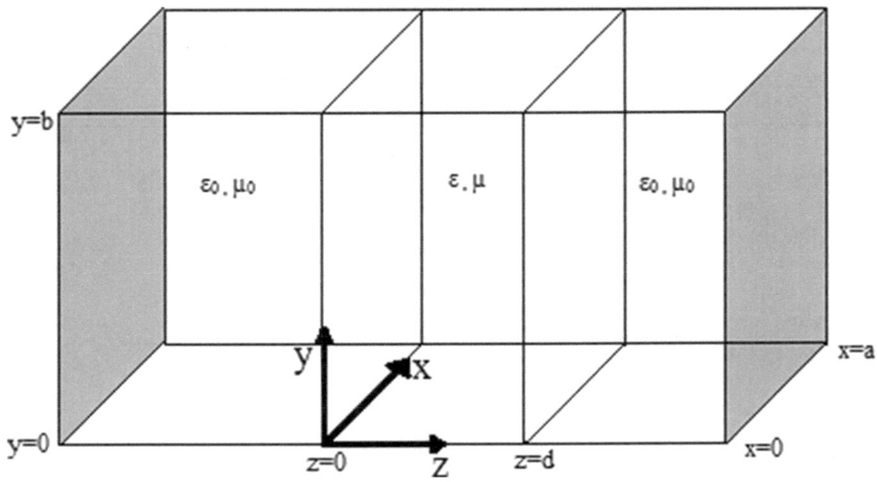

Fig. 3.3 Waveguide system with the presence of material

working environment, with proper estimated measurement accuracy. For material characterization, the material under test (MUT) is positioned into the system for the extraction of known values of permittivity, ε and permeability, μ using a precise algorithm. In this process, classic 'Nicolson–Ross–Weir' (NRW) algorithm (Fenner et al. 2012) is used for obtaining the reflection and transmission coefficients of the material filled inside the waveguide. Figure 3.3 shows the configuration used in NRW method. The sample with unknown properties is placed in the waveguide system, filling the region $0 \leq z \leq d$. S-parameters are measured from the vector network analyzer (VNA) and mathematically altered to obtain S-parameters at the illustration plane. These S-parameters are used to obtain the sample propagation constant, β and interfacial reflection coefficient, Γ. These in turn are used for obtaining ε and μ. Interfacial reflection coefficient can be expressed as follows:

$$\Gamma = \frac{1 - V_1 V_2}{V_1 - V_2} \pm \sqrt{\left(\frac{1 - V_1 V_2}{V_1 - V_2}\right)^2 - 1} \tag{3.19}$$

where $V_1 = S_{21} + S_{11}$ and $V_2 = S_{21} - S_{11}$, the sign is chosen appropriately such that $|\Gamma| \leq 1$. The one way-propagation factor of wave roving through the illustration section can be articulated in terms of Γ as

$$P = \mathrm{e}^{-\gamma d} = \frac{V_1 - \Gamma}{1 - \Gamma V_1}, \tag{3.20}$$

where $\gamma = (\ln P - j2n\pi)/(-d)$ and n is an integer whose value depends on the length of the sample. The permittivity and permeability can be expressed as

$$\mu = \mu_o \frac{1+\Gamma}{1-\Gamma} \frac{\gamma}{j\beta_o}, \quad \varepsilon = \varepsilon_o \left(\frac{\gamma \beta_o}{jk_o^2} \frac{1-\Gamma}{1+\Gamma} + \frac{\mu_o}{\mu} \frac{k_c^2}{k_o^2} \right) \tag{3.21}$$

where $k_o = \omega\sqrt{\mu_o \varepsilon_o}$, $\beta_o = \sqrt{k_o^2 - k_c^2}$ and $k_c = \pi/a$

The designing of the substitute should be done precisely with much care so that the measured S-parameter gives appropriate values of permittivity and permeability. Because of the non-magnetic nature of the substitute, it is advisable to evaluate the material properties in lossless condition.

The waveguide standard must be built using materials that are commonly available in abundance and with which standard can be effortlessly fabricated. Hence, the choice of sample must comprise solely non-magnetic metallic element as well as flawlessly conducting material. Figure 3.4 shows a simple structure consisting of two indistinguishable rectangular apertures of width $\Delta^l = \Delta^r$ detached by a space of Δ^s. This simple structure has several advantages compared to a complex structure. First, this structure can be easily engineered with the assistance of metallic sheets. Second, mode-matching technique can be used for analysis of the sample rapidly and with superior accuracy. Third, if some other standard materials, different from that described after optimized design, are used, then mode-matching can be utilized for resolving the theoretical properties of the artificial sample and therefore can be used for analysis of the material properties.

The sample is placed in the measurement setup as shown in Fig. 3.4. The NWR method presumes the sample to be isotropic and homogenous as a result of which the following considerations are taken into account: $S_{11} = S_{22}$ and $S_{12} = S_{21}$. The waveguide model ought to be in equilibrium in the longitudinal direction, which

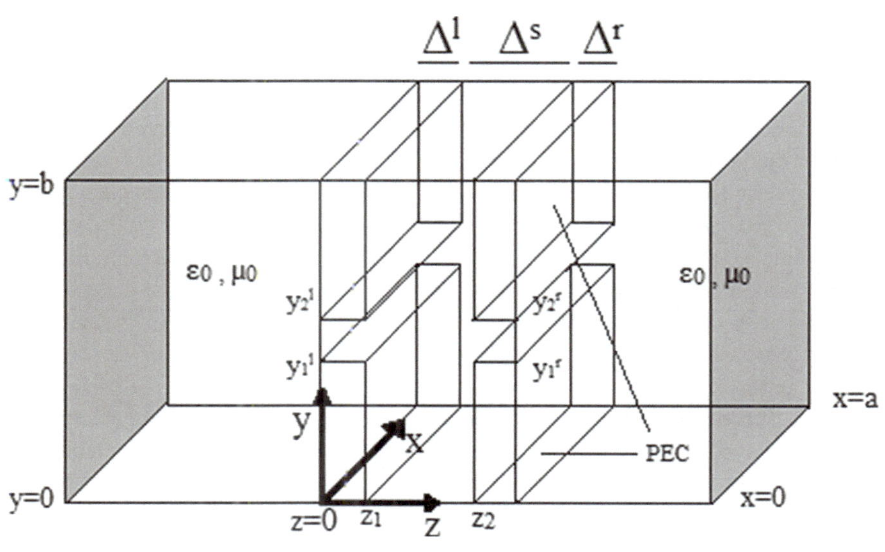

Fig. 3.4 Waveguide system in presence of verification standard

helps in getting identical results and in addition, if lossless materials are used for construction then purely real permittivity and permeability values can be extracted with NRW method. The permittivity and permeability are frequency dependent if sample is inhomogeneous along the longitude.

Equal permittivity and permeability are a desirable property required for a material in the construction of microwave structure applications in the range of 3–10 GHz. Hence, this design issued for tuning using a Genetic algorithm (GA) to obtain the desired sample. The GA differed the vertical opening locations (locations $y_1^l = y_1^r$ and $y_2^l = y_2^r$ as in Fig. 3.4), horizontal opening locations and the width of the apertures and waveguide bar. The simulation software HFSS is used for the computation of S_{11} and S_{21}. Optimizer configuration is a crossover of 2-point and 3-point boundary with a progressing single bit transformation. The crossover points were nominated arbitrarily. The geometry of the waveguide sample is created using Matlab by selecting randomly a population of 100 diverse binary strings. This structure is then simulated in HFSS by exporting the Matlab file to HFSS.

The waveguide used in this method is a standard S-band WR-284 having the dimension of 72.136 × 34.036 mm. The working band of frequency is 2.6–3.95 GHz. For the simplification of fabrication process, the width of each aperture and bar for GA optimization are limited to the values of United States standard brass stock that is multiples of 1/16 in. or 1.5875 mm. Apertures are allocated identical openings for symmetric confirmation standards.

The parameters extracted by means of NRW method from the simulated S-parameter has zero as the imaginary. The accurate values for the material parameters created by the optimized waveguide may be initiated by the application of mode-matching technique. An accuracy up to 5 digits in each extracted parameters is assured, by selecting the value of N appropriately, for producing the mode-matching result. The obtained result of mode-matching is analogous to the one extracted from HFSS. Computing the S-parameters in HFSS takes 60 times more time match up to mode-matching on a 3.5 GHz Intel core processor with 24 GB of RAM. Thus, Monte Carlo error can be analysed by the faster mode-matching approach with the purpose of assessing the sensitivity of the fabricated standard to the geometry ambiguities.

According to the geometry shown in Fig. 3.4, the sample materials placed at the position $z = 0$ and $z = z_2$ having width of Δ^l and Δ^r, respectively. The thickness of the sample is allowed to be different so as to avoid the effect of manufacturing errors. Sample width is same as blank waveguide. The length of the waveguide is assumed to be sufficient, such that even though a full spectrum of greater order modes is created at the sample, only a TE_{10} mode is acknowledged at the determining port.

It is assumed that a TE_{10} mode is incident from the conducting extension $z < 0$. An infinite spectrum of modes is generated in each of the waveguide segments due to the connections of fields with the sample at $z = 0$. Since incident field is symmetric, only TE_{1v} and TM_{1v} modes are excited with non-zero amplitudes. The transverse field of every waveguide segment can be represented in modal series where forward and reverse waves are signified. A number of modes used in each region are identical. On application of the boundary conditions on the transverse fields between the full waveguide and the sample regions, the modal amplitude can be determined. The modal amplitudes produce a system of functional equations. These functional equations can be changed to a system of linear equations by the application of suitable operators. The system of linear equations can be written as $8N \times 8N$ portioned matrix. The integral computation is done in closed form, and the matrix access is repetitive or zero. This heads towards the efficiency of the mode-matching approach by allowing the matrix to be filled rapidly.

The S-parameters are written in terms of the modal coefficients as follows:

$$S_{11} = \frac{a_1^-}{a_1^+} \tag{3.22}$$

$$S_{21} = \frac{f_1^+}{a_1^+} \tag{3.23}$$

where a_1^+ represents the amplitude of the incident TE_{10} mode in transmitting section, a_1^- represents the amplitude of the reflected TE_{10} mode transmitting section and f_1^+ represents the amplitude of the forward travelling TE_{10} mode in the receiving section.

It can be concluded that, in this method (Crowgey et al. 2015) a functional authentication of sample material is initiated using a waveguide standard for material characterization in the microwave spectrum. The requirement of the sample is that it is gradually shifting, conventional and precisely replicating magnetic characteristics at microwave frequencies. The material used for fabrication of the standard are all metal as it is easy for fabrication and uncomplicated for designing using mode-matching approach. Nicolson–Ross–Weir method is used for obtaining relative permittivity and permeability in the *S*-band using specific measurements. The optimization of the sample is done using a genetic algorithm so as to reduce the sensibility of the extracted parameters towards the geometric parameters as well as for reduction of tolerance.

3.1.4 An Estimated Method for Resolving Accuracy Problem for Permittivity and Permeability Values' Near λ/2 Resonances in Transmission/Reflection Technique

Detailed information about the properties of a material such as its electric permittivity, magnetic permeability as well as electric and magnetic losses are essential parameters for any scientific research on materials which can be used for engineering application.

Solids form the chief class of materials whose properties are tested over broadband frequency range. In broadband transmission, the method which is predominantly used for extracting the material parameters is transmission/reflection method. The conventional transmission/reflection method fails at resolving two major intrinsic issues. The first being explicit or non-iterative T/R methods, where calculation of natural logarithm containing measured transmission and reflection data is done and if the material being tested is dispersive, the branch cut becomes abstruse. The other results from the standing waves arising within a low loss sample. This happens due to impedance mismatch of transmission line with that of the air and it occurs on both sides of the sample faces. The dimensions of the sample faces are integer multiple of half wave length. This geometric resonance is called as 'Fabry-Perot' resonance as explained in Chalapat et al. (2009), Qi et al. (2010), Liu et al. (2011).

The obtained permittivities and permeabilities become oscillatory resulting in accuracy degradation of material characterizations. This occurs due to the half wave ($\lambda/2$) resonances in the transmission/reflection coefficients of the low loss samples. This issues are discussed in Chalapat et al. (2009) Boughriet et al. (1997), Hasar and Westgate (2009). The artifacts for extracted permittivity have been resolved at $\lambda/2$ resonance using the proposed method.

Nicolson–Ross–Weir (NRW) method was reformulated (Boughriet et al. 1997) for a low loss material with the assumption that $\mu_r = 1$, which led to smoothing of permittivity data at resonance. An exclusive amplitude-based method was developed by Hasar (2008), Hasar and Westgate (2009). This method is used for determining permittivities of low loss materials with the assumption that, qualms in phase of reflection are overriding at resonance. Baker-Jarvis technique Baker-Jarvis et al. (1990, 1993) was modified by Chalapat et al. (2009) by inclusion of a reference-plane-invariant method. This method helps in the reduction of errors for permittivity as compared to original T/R methods. But in all these above methods, only the value of permittivity is checked and not permeability.

The method proposed in this technique overcomes the demerit of previous methods as it helps in the determination of both permittivity and permeability of a low loss dielectric material at around half wave resonance. First, a demarcation of the $\lambda/2$ resonance issue is made for original T/R method. For the proposed method, a start point is offered in consort with a spontaneous description regarding the behaviour of the input impedance for sample-loaded transmission line at the resonance. The characteristic impedance of the sample-filled section is calculated using first-order regression coefficients for input impedance with an assumption of sample material being non-dispersive. The sole assumption taken in this technique is that the characteristic impedance of the material occupied segment of the transmission line has a flat response at frequencies near the half-wavelength resonance.

3.1.4.1 Problems in Conventional T/R Methods

Here is a brief review about the concerns that stand up in conventional T/R methods, frequently discussed as Nicolson–Ross–Weir NRW) method (Weir 1974; Nicolson and Ross 1970) Baker-Jarvis (BJ) iterative method (Baker-Jarvis et al. 1990, 1993), etc.

In this methods, complex permittivity and permeability of a sample positioned in transmission line are calculated from the obtained line of *S*-parameter data for a transmission line packed with test material is presented in the form of a diagram in Fig. 3.5. The different parameters of the test material are represented by refractive index as n, relative wave impedance as ς, absolute permittivity as ε, absolute permeability as μ, propagation constant as γ and characteristic impedance as Z.

For the implementation of NRW method, the measured *S*-parameters are as follows:

$$S_{11} = S_{22} = \frac{\Gamma(1 - t^2)}{1 - \Gamma^2 t^2} \tag{3.24}$$

$$S_{21} = S_{12} = \frac{t(1 - \Gamma^2)}{1 - \Gamma^2 t^2} \tag{3.25}$$

where t is the transmission coefficient and Γ is the reflection coefficient of the interface.

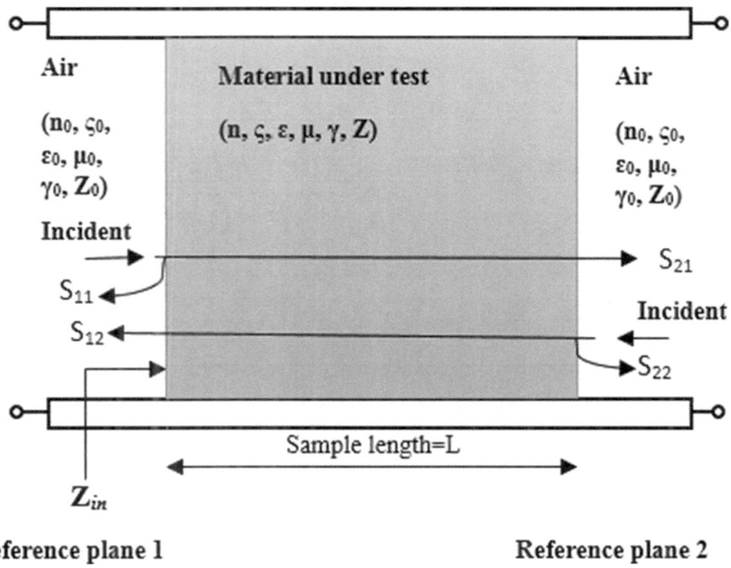

Fig. 3.5 Diagrammatic representation of *S*-parameter and input impedance measurement of sample-loaded transmission line

Hence, the material parameters are connected as

$$t = \exp(-\gamma L) \tag{3.26}$$

$$\Gamma = \frac{\frac{\mu}{\gamma} - \frac{\mu_0}{\gamma_0}}{\frac{\mu}{\gamma} + \frac{\mu_0}{\gamma_0}} \tag{3.27}$$

where L represents the sample length.

The propagation constant, γ is written as

$$\gamma = j\sqrt{\frac{\omega^2 \varepsilon_r \mu_r}{c^2} - \left(\frac{2\pi}{\lambda_c}\right)^2} \tag{3.28}$$

where j is the imaginary unit, ε_r is the relative permittivity, μ_r is the relative permeability, c is the speed of light, ω is the angular frequency and λ_0 is the cut-off wavelength of rectangular waveguide.

According to NRW technique, precise computation of Γ and t are carried out with the help of measured value of S_{11} and S_{21} and are expressed as

$$\Gamma = X \pm \sqrt{X^2 - 1} \tag{3.29}$$

$$X = \frac{\left(S_{11}^2 - S_{21}^2\right) + 1}{2S_{11}} \tag{3.30}$$

$$t = \frac{(S_{11} + S_{21}) - \Gamma}{1 + (S_{11} + S_{21})\Gamma} \tag{3.31}$$

Using Eqs. (3.29)–(3.31) we get

$$\mu_r = \frac{1 + \Gamma}{(1 - \Gamma)\Lambda\sqrt{(1/\lambda_0^2) - (1/\lambda_c^2)}} \tag{3.32}$$

$$\varepsilon_r = \frac{\lambda_0^2}{\mu_r}\left(\frac{1}{\Lambda^2} + \frac{1}{\lambda_c^2}\right) \tag{3.33}$$

λ_0 is the free-space wavelength and Λ is the guided waveguide wavelength represented as

$$\frac{1}{\Lambda^2} = -\left[\frac{1}{2\pi L}\ln\left(\frac{1}{t}\right)\right]^2 \tag{3.34}$$

The sample under test is passive so the choice in sign for Eq. (3.29) should be such that $|\Gamma| \leq 1$. Increasing the sample length beyond the half wave resonance may cause multiple values as in Eq. (3.34) that is the logarithm. Depending on the smoothness of the measured group delay at desired frequencies, the selection of the correct branch of the algorithm (Weir 1974) can be made. If the sample of material has dispersive characteristics, the correct branch cut-off would remain uncertain. However, this ambiguity is not considered in this work and the test method remains non-dispersive.

In BJ method, the solution is obtained iteratively by reformulating Eqs. (3.24) and (3.25) to obtain ε_r and μ_r.

$$S_{11}S_{22} - S_{21}S_{12} = \exp[-2\gamma_0(L_{\mathrm{air}} - L)]\frac{\Gamma^2 - t^2}{1 - \Gamma^2 t^2} \quad (3.35)$$

$$\frac{S_{21} + S_{12}}{2} = \exp[-\gamma_0(L_{\mathrm{air}} - L)]\frac{t(1 - \Gamma^2)}{1 - \Gamma^2 t^2} \quad (3.36)$$

Here, L_{air} represents length of the fixture including the transmission line such that $L_{\mathrm{air}} \geq L$

Unlike NRW method where the reference plane is in front and back faces, the reference plane for BJ iteration method need not necessarily be at the back and front faces of the sample. Equations (3.35) and (3.36) show that the determinant of scattering matrix and arithmetic mean of the transmission coefficients are function of $L_{\mathrm{air}} - L$ and not only L. Hence, BJ iterative method also termed as the reference-plane invariant.

Considering TEM case for ease, the equations are rewritten in alternative ways for the clarification of inherent problems in conventional T/R methods.

The refractive index n of the material can be formulated from Eqs. (3.33) and (3.34) considering $\lambda_c = \infty$ for Eq. (3.33)

$$n = n' - jn'' = \pm\sqrt{\varepsilon_r\mu_r} = \pm j\frac{\lambda_0}{2\pi t}\ln\left(\frac{1}{t}\right) \quad (3.37)$$

Substituting the value of X from Eq. (3.30) to (3.29)

$$\Gamma^2 - \Gamma\left[\frac{1 - \left(S_{21}^2 - S_{11}^2\right)}{S_{11}}\right] + 1 = 0 \quad (3.38)$$

Reflection coefficient Γ is related to relative wave impedance as follows:

$$\Gamma = (\varsigma - 1)/(\varsigma + 1)$$

Substituting the above equation to Eq. (3.38) we get

$$\varsigma = \varsigma' + j\varsigma'' = \pm\sqrt{\frac{\mu_r}{\varepsilon_r}} = \pm\sqrt{\frac{(1 + S_{11})^2 - S_{21}^2}{(1 - S_{11})^2 - S_{21}^2}} \quad (3.39)$$

The condition for selection of signs in Eqs. (3.37) and (3.39) considering the passivity constraints, are $n'' \geq 0$ and $\varsigma'' \geq 0$, respectively.

Hence, permittivity and permeability can be calculated from $\varepsilon_r(= \varepsilon_r' - j\varepsilon_r'') = n/\varsigma$ and $\mu_r(= \mu_r' - j\mu_r'') = n \cdot \varsigma$, respectively, by substituting the value of n and ς from Eqs. (3.37) and (3.39). Equation (3.39) offers an appropriate way of how accuracy is lost at around the half-wavelength resonance frequencies.

In this section, it is discussed that if T/R method is being derived from Eqs. (3.24) and (3.25) then its permittivities and permeabilities will be contaminated with artifacts at around $\lambda/2$ resonance. This situation sufficiently explains the constraints related to a conventional T/R method, and hence it is difficult to achieve material parameter for the low loss sample at each frequency point near $\lambda/2$ resonance. If a reasonable wave impedance is developed around $\lambda/2$ resonance then there is a possibility to remove the artifacts near the resonance.

3.1.4.2 Input Impedance of Sample-Loaded Transmission Line

Input impedance as a function of refractive index and wave impedance of the material under test in the transmission line are examined in this section. Considering the transmission line loaded with material having n and ς as shown in Fig. 3.5, the following assumptions are made. The first reference plane, i.e. reference plane 1 is coinciding with the front face of the sample but the connection between back face and reference plane 2 is kept arbitrary, sample-loaded transmission line is introduced between the transmission lines having characteristic impedance of Z_0 connecting to VNA. Input impedance Z_{in} when seen from reference plane 1 is written as

$$Z_{in} = R + jX = Z\frac{Z_0 + Z\tan h(\gamma L)}{Z + Z_0 \tan h(\gamma L)} \tag{3.40}$$

where R represents the real resistive part of input impedance, X represents the imaginary reactive part of input impedance, γ represents the propagation constant and Z represents the characteristic impedance of the sample-filled section .

$$\begin{aligned} \gamma &= jk_0 n (\text{coaxial transmission line}) \\ \Rightarrow \gamma &= j\sqrt{(k_0 n)^2 - \left(\frac{2\pi}{\lambda_c}\right)^2} (\text{rectangular waveguide}) \end{aligned} \tag{3.41}$$

$$Z = Z' + jZ'' \Rightarrow Z = \frac{\varsigma\varsigma_0}{2\pi}\ln\left(\frac{b}{a}\right) \text{ (coaxial transmission line)}$$
$$\Rightarrow Z = \frac{j\omega\mu_r\mu_0}{\gamma} = \frac{\omega\mu_0\mu_r}{\sqrt{(k_0 n)^2 - (2\pi/\lambda_c)^2}} \text{ (rectangular waveguide)} \tag{3.42}$$

Again, k_0 is the wavenumber, ς_0 is the absolute wave impedance (377 Ω) in air, a and b are the radii of inner and outer conductors of the coaxial transmission line.

3.1.4.3 New Approach for Reducing Artifacts Near λ/2 Resonance

For the calculation of characteristic impedance Z of the sample-filled area, an approximation of the slope of real and imaginary parts of R and X of the already measured input impedance Z_{in} is taken with the assumption that Z is constant around $\lambda/2$ resonance. An illustration for fitting the real and imaginary R and X of the input impedance Z_{in} of the sample-loaded transmission line for some first order regression is shown in Fig. 3.6.

Discussing Fig. 3.6a, it can be seen from the diagram that the resonant frequency band has been divided into two region being $f_A - f_0$ and second being $f_0 - f_B$. Here, f_0 is the centre frequency, f_A is the start frequency at resonance and f_B is the stop frequency at resonance.

The value of f_0 should be chosen such that R is maximal or minimal at that frequency and $X = 0$ at that same time. The regression lines expression, for the different regions, for approximation of measured R and X with first order are

Region I $R = a_1 f + a_2$ and $X = b_1 f + b_2$

Region II $R = c_1 f + c_2$ and $X = d_1 f + d_2$

Here, a_1, b_1, c_1, d_1 are the regression coefficients represented as

$$a_1 = \frac{R(f_0) - R(f_A)}{f_0 - f_A} \tag{3.43}$$

$$b_1 = \frac{X(f_0) - X(f_A)}{f_0 - f_A} \tag{3.44}$$

$$c_1 = \frac{R(f_B) - R(f_0)}{f_B - f_0} \tag{3.45}$$

$$d_1 = \frac{X(f_B) - X(f_0)}{f_B - f_0} \tag{3.46}$$

The transmission line characteristic impedance Z in region I can be promptly solved using Eqs. (3.43) and (3.44) and using Eqs. (3.45) and (3.46), Z can be solved for region II. If frequencies f_0, f_A and f_B are very high then the coefficients a_1, b_1, c_1 and d_1 possibly will be very small causing a degradation on accuracy of Z. Hence GHz range frequencies are used to avoid such circumstances.

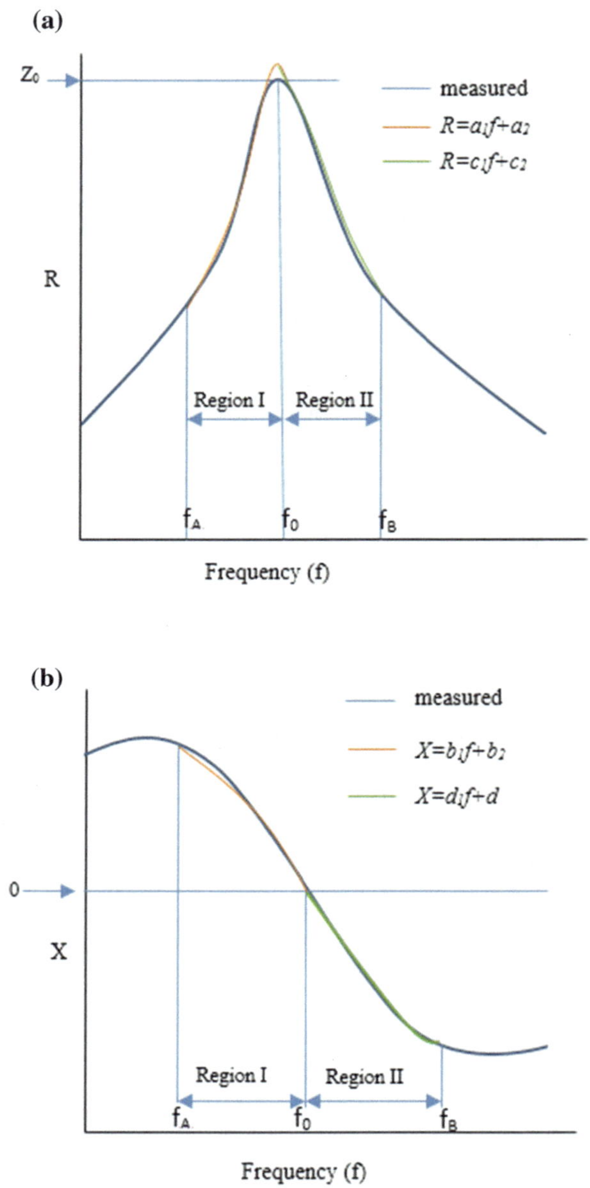

Fig. 3.6 Illustration of regression for **a** R **b** X of input impedance Z_{in}

The wave impedance for a coaxial transmission line can be obtained with the data of $Z = (Z' + jZ'')$:

$$\varsigma = \frac{2\pi Z}{\varsigma_0 \ln\left(\frac{b}{a}\right)} \tag{3.47}$$

Permittivity and permeability can be obtained from Eqs. (3.47) and (3.37) with ς and n, respectively,

that is $\varepsilon_r = n/\varsigma$ and $\mu_r = n \cdot \varsigma$.

For a rectangular waveguide, relative permeability is calculated with data obtained from Z as

$$\mu_r = \frac{\gamma Z}{j\omega\mu_0} = \frac{\sqrt{(k_0 n)^2 - (2\pi/\lambda_c)^2}}{\omega\mu_0} Z \tag{3.48}$$

Permittivity can also be written as $\varepsilon_r = n^2/\mu_r$. Now substituting the value of μ from Eq. (3.48) and value of n from Eq. (3.33) along with (3.34) $\left(n = \pm\sqrt{\varepsilon_r \mu_r} = \pm\lambda_0\sqrt{1/\Gamma^2 + 1/\lambda^2}\right)$.

For obtaining flat Z, an assumption is taken according to which the sample should have non-dispersive material parameters near resonance. Further, this assumption guarantees that branch cut ambiguity is avoided when n is calculated from Eq. (3.37).

Figure 3.7 shows a flow diagram for the computational procedure of both coaxial transmission line and rectangular waveguide. The refractive index n is obtained from the measured S-parameter and input impedance Z_{in} similarly as is done in ordinary T/R method. After which regression coefficients a_1, b_1, c_1, d_1 are calculated from Z_{in} so as to find the characteristics impedance Z of the loaded section.

The only difference between coaxial transmission line and rectangular waveguide is that for coaxial line wave impedance ς is computed from Z while for rectangular waveguide, relative permittivity is obtained immediately and the rest follows the same for both the transmission lines.

In conventional T/R method, considering NRW method two equations that of S_{11} and S_{21} are required for solving the unknown permittivity and permeability and for that of BJ method all four of S-parameters, i.e. S_{11}, S_{12}, S_{21}, S_{22} are required for the calculation of unknowns. In this work initially, Eq. (3.25) was used for measurement of S_{21} and Eq. (3.40) for input impedance Z_{in} instead of using Eq. (3.24) for S_{11}. Though the same result for c was obtained as that of conventional T/R method. Furthermore, using equations for resonant frequency f_0 and quality factor Q at $\lambda/2$ resonance in place of Eq. (3.24) results in very low value of Q for various material samples for obtaining accurate ε_r and μ_r from calculated f_0 and Q.

This approach is considerably more accurate compared to NRW method and BJ method. This approach can be used in any dielectric material as long as it possesses flat material properties at $\lambda/2$ resonant frequency. Importantly, the accuracy may also vary depending on the values of f_A and f_B.

Fig. 3.7 Flow chart of computational procedure

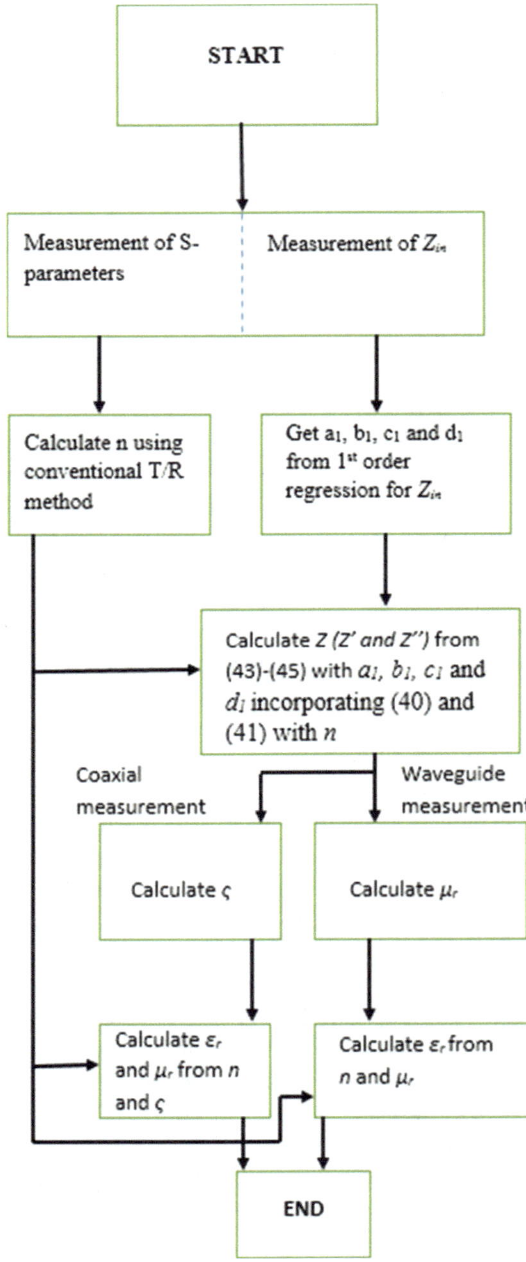

3.2 Stripline Methods

In the design process of metamaterial-based structures, the EM characterization method for the selected structure plays an important role. The EM material characterization of metamaterials using waveguides, resonant cavities and free-space systems are expensive regarding fabrication aspects. Further, these techniques have limited bandwidths and often need different test fixtures for different frequency bands. In this regard, stripline-based methods have obvious advantages due to its large operational bandwidth and easiness of fabrication (simple test fixtures).

3.2.1 *Microstrip Line Method*

In this technique, the metamaterial sample is used as the substrate for microstrip line (Fig. 3.8). Here, the characteristic impedance and propagation constants are a function of the permittivity of the substrate. For the test fixture considered here, the characteristic impedance and propagation constant are given by (Boybay et al. 2010).

$$Z = \frac{\eta_o}{2} \frac{\sqrt{1/\varepsilon}}{\frac{w}{2h} + \frac{1}{\pi}\ln[4] + \frac{\varepsilon+1}{2\pi\varepsilon}\ln\left[\frac{e\pi}{2}\left(\frac{w}{2h}+0.94\right)\right] + \frac{\varepsilon-1}{2\pi\varepsilon^2}\ln\left[\frac{e\pi^2}{16}\right]} \tag{3.49}$$

$$\beta = \omega\sqrt{\varepsilon_o\mu_o}\frac{Z_o}{Z} \tag{3.50}$$

The characteristic impedance of substrate is given by

$$Z_o = \frac{\eta_o}{2} \frac{1}{\frac{w}{2h} + \frac{1}{\pi}\ln\left[2e\pi\left(\frac{w}{2h}+0.94\right)\right]} \tag{3.51}$$

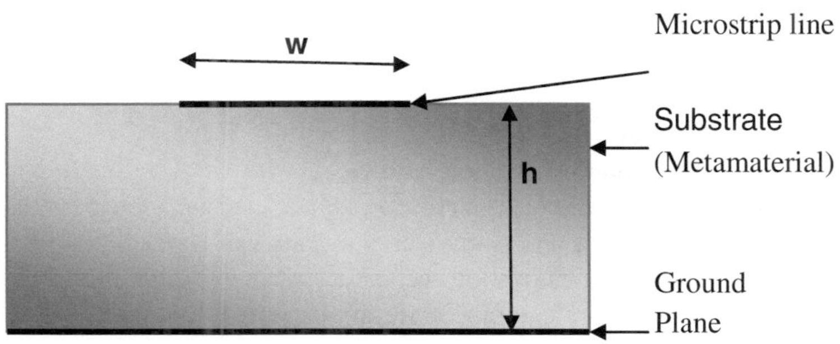

Fig. 3.8 Cross section of microstrip line structure

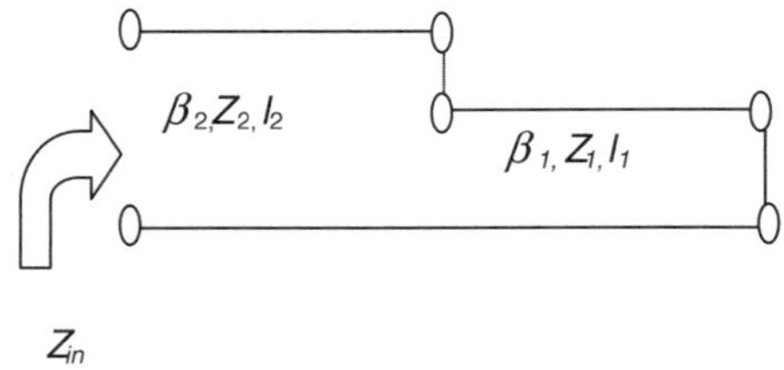

Fig. 3.9 Transmission line representation of the structure

where ε is the relative permittivity of substrate, *e* is the Euler's number, η_0 is the free-space wave impedance, *w* is the length of microstrip line, *h* is the height of substrate, *β* is the propagation constant, *Z* is the characteristic impedance and Z_o is the characteristic impedance of free space.

It is noted that the propagation constant in the metamaterial region is imaginary. In order to have a physically consistent system, the sign for propagation constant must be selected carefully. In the formulae presented above, an assumption of $w \geq 2h$ has been made.

A microstrip line having ε-negative substrate is terminated by a short circuit and is excited by a microstrip line with ε-positive substrate. Figure 3.9 represents the transmission line representation of the above-mentioned configuration.

Except for β_1 and Z_1 (function of permittivity of the substrate) of the first microstrip transmission line, all the other parameters are assumed to be known. For the second transmission line, it is assumed to have a substrate permittivity of 1. Further, it is assumed that both transmission lines have the same ratio of w/h. Equations (3.49)–(3.51) are valid for substrates with negative permittivity. The formulation shown above can be used only when w/h ratio is >3. In case of narrow strips, the formulation is valid only for substrates with low values of permittivity. This method is a wide band characterization technique for metamaterials and also cost effective.

3.2.2 Measurement Technique for Microstrip Circuit Applications

A measurement technique based on *S*-shaped resonator and feedback transmission line is explained in this method (Nesimogluand Sabah 2014) (Fig. 3.10).

The uniqueness of this method is listed below:

- Realization of metamaterial is flexible.
- Diminution in the count of periodic unit cells.
- For the achievement of hybrid microwave circuits, metamaterials with surface mount microwave active and passive components can be used.
- New experimental characterization techniques of metamaterial with the aid of microstrip and laboratory equipment are implemented.
- Simplicity in the use of countless PCB and microwave applications.

Fig. 3.10 Metamaterial structure. *Top* front view, *bottom* back view

The effective constitutive parameters are retrieved and analysed. This method can be used for the attainment of new metamaterialsthat can be utilized in numerous microstrip-based microwave applications.

For the realization of artificial metamaterials possessing eccentric properties, *S*-shaped resonator (SSR) and feedback transmission line (FTL) are united mutually. These two in particular are combined for their superior properties. In this structure, the substrate used is FR4 having a thickness of 1.6 mm, dielectric constant of 4.4 and loss tangent of 0.02. The SSR and FTL are made of copper wires having conductivity of 5.8×10^7 S/m and thickness of 0.036 mm. The structure is designed such that SSR comes on one face of the substrate and FTL comes on the opposite face. The dimensions of both the structures are calculated accurately in an attempt to acquire a characteristic impedance of 50 Ω. This particular standard of 50 Ω is considered because it is compatible with laboratory test equipment.

A standard simulation tool based on finite integration technique is used here. Appropriate boundary conditions like magnetic, electric and open boundary conditions are set before simulation is started. All these combined together gives the S-parameter of the structure. For this structure, the transmission is approximately zero whereas reflection is maximum at the resonance frequency.

At certain frequencies, there occurs a jump in the transmission and reflection phases, this indicating its proximity to the resonant frequency which in turn indicates the characteristics of metamaterial resonance. The electromagnetic features of the designed structure like permittivity and permeability are extracted from its *S*-parameters. All these parameters are frequency dependent and gratifies the requests of causality.

In a nut shell, an S-shaped metamaterial with feedback transmission line is constructed maintaining an appropriate width of the two structures so as to maintain a characteristic impedance of 50 Ω. A ground frame is inserted and measurements are taken using standard test equipment. SMA connectors are used for the excitations as well as to procure the *S*-parameters of the structure from the side opposite to that used for excitation. The *S*-parameters for the fabricated structure are obtained using VNA and in case of simulations, using software. The *S*-parameters can be plotted using the obtained result.

3.2.3 Stripline Fixture Method

A method based on stripline fixture has been reported for the characterization of all types of metamaterialsstructures including single-negative and double negative media (Yousefi et al. 2011). The sample under test is used as the substrate of a stripline structure and the complex permittivity and permeability of the sample are extracted from the measured *S*-parameters. Figure 3.8 shows the two-port stripline fixture. The substrate of the stripline comprises of three parts: two double positive dielectric having known constitutive parameters at the two sides next to the excitation ports, and the metamaterial that has to be characterized is placed in the middle. By measuring the *S*-parameters of the two-port stripline fixture, permittivity and permeability of the metamaterial under test can be extracted. Here, the y-component of the magnetic field and the x-component of the electric field are the dominant field components in the stripline structure as shown in Fig. 3.8.

The theoretical basis of this method is similar to that of free-space approach since the two methods use reflection and transmission of waves from the metamaterial sample for extracting the constitutive parameters. Three transmission line sections are used for representing the three regions of the stripline fixture shown in Fig. 3.9 and the behaviour of the field in the substrate is analysed by transmission line model.

The voltage and current in all the three regions are represented by

$$\text{Region I}: \quad V_{\mathrm{I}} = \mathrm{e}^{-jk_{\mathrm{d}}z} + \Gamma \mathrm{e}^{jk_{\mathrm{d}}z} \tag{3.52}$$

$$I_{\mathrm{I}} = \frac{1}{Z_{\mathrm{d}}}\left(\mathrm{e}^{-jk_{\mathrm{d}}z} - \Gamma \mathrm{e}^{jk_{\mathrm{d}}z}\right) \tag{3.53}$$

$$\text{Region II}: \quad V_{\mathrm{II}} = C\mathrm{e}^{-jk_{\mathrm{m}}z} + D\mathrm{e}^{jk_{\mathrm{m}}z} \tag{3.54}$$

$$I_{\mathrm{II}} = \frac{1}{Z_{m}}\left(C\mathrm{e}^{-jk_{\mathrm{m}}z} - D\mathrm{e}^{jk_{\mathrm{m}}z}\right) \tag{3.55}$$

$$\text{Region III}: \quad V_{\mathrm{III}} = T\mathrm{e}^{-jk_{\mathrm{d}}(Z-t)} \tag{3.56}$$

$$I_{\mathrm{III}} = \frac{1}{Z_{d}}\mathrm{e}^{-jk_{\mathrm{d}}(Z-t)} \tag{3.57}$$

where k_{d} is the propagation constant and Z_{d} is the characteristic impedance in the known dielectric (Regions I, and III in Fig. 3.11); k_{m} is the propagation constant and Z_{m} is the characteristic impedance in the unknown metamaterial sample, respectively (Region II in Fig. 3.11).

As the dielectric in Regions I and III are nonmagnetic and isotropic, k_{d} and Z_{d} can be written as

$$k_{\mathrm{d}} = \omega\sqrt{\varepsilon_o \varepsilon_{\mathrm{d}} \mu_o} \tag{3.58}$$

$$W_e = \begin{cases} W & W/h > 0.35 \\ W - h(0.35 - W/h)^2 & W/h < 0.35 \end{cases} \tag{3.59}$$

$$Z_{\mathrm{d}} = 30\pi\sqrt{\frac{\mu_o}{\varepsilon_o \varepsilon_{\mathrm{d}}}}\left[\frac{h}{W_e + 0.441h}\right] \tag{3.60}$$

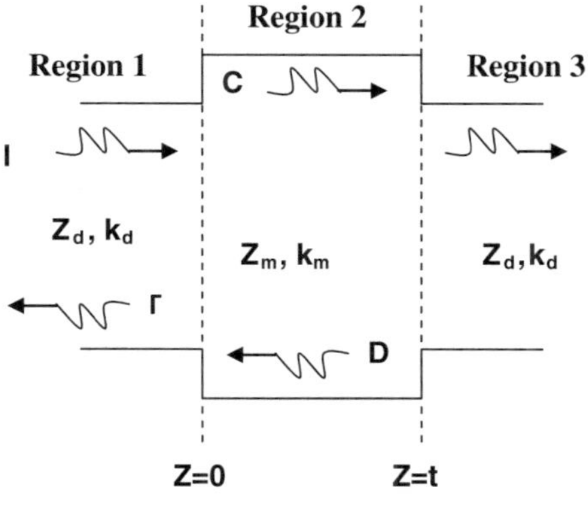

Fig. 3.11 Transmission line model of the fixture

where W is the width of the stripline, h is the total height of the substrate, ε_d is the relative permittivity of the known dielectric. The direction of the E and H fields should be considered when deriving k_m and Z_m, as the metamaterial sample is generally anisotropic. As shown in Fig. 3.12, the dominant E component is in x-direction and the dominant H component is in y-direction, for the stripline topology. Then k_m and Z_m can be obtained as

$$k_m = \omega\sqrt{\varepsilon_o\mu_o\varepsilon_x\mu_x} \tag{3.61}$$

$$W_e = \begin{cases} W & W/h > 0.35 \\ W - h(0.35 - W/h)^2 & W/h < 0.35 \end{cases} \tag{3.62}$$

$$Z_m = 30\pi\sqrt{\frac{\mu_o\mu_y}{\varepsilon_o\varepsilon_x}}\left[\frac{h}{W_e + 0.441h}\right] \tag{3.63}$$

where μ_y is the permeability in y-direction and ε_x is the permittivity in x-direction.

Equations (3.53)–(3.58) are solved by applying the boundary conditions at $z = 0$ and $z = t$ so as to express the permittivity and permeability of the metamaterial in terms of the measured S-parameters. This yields the following relationships:

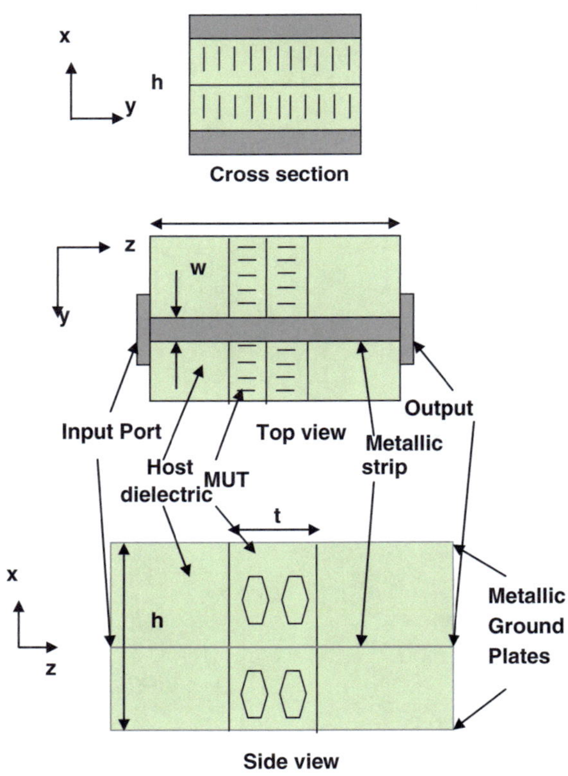

Fig. 3.12 Stripline fixture used for extraction of the permittivity and permeability of the metamaterial media

$$\mu_y = \frac{Z_m k_m}{Z_d k_d}, \quad \varepsilon_x = \frac{\varepsilon_d k_m Z_d}{Z_m k_d} \tag{3.64}$$

$$Z_m = \pm Z_d \sqrt{\frac{(1+\Gamma)^2 - T^2 e^{-j2k_d t}}{(1-\Gamma)^2 - T^2 e^{-j2k_d t}}} \tag{3.65}$$

$$\Gamma = S_{11} e^{-j2k_d d}, \quad T = S_{21} e^{-j2k_d d} \tag{3.66}$$

$$X = \frac{1 - \Gamma^2 + T^2 e^{-j2k_d t}}{2T e^{-j2k_d t}} \tag{3.67}$$

$$e^{-jk_m t} = X \pm j\sqrt{1 - X^2} \tag{3.68}$$

The equations presented for Z_m (Eq. 3.64) and Z_d (Eq. 3.61) are approximate formulas. Since the width of the stripline W and the height of the substrate h are the same for the dielectric part and metamaterial part, the bracketed expressions in Eqs. (3.64) and (3.61) are identical. The bracketed parts of the expressions in Z_m and Z_d cancels out, when calculating the extracted parameters, μ_y and ε_x. As a result, the approximations embedded in Z_m and Z_d relating line width and substrate height will not affect the extracted parameters, μ_y and ε_x, hence the accuracy of the method.

The method presented above is inexpensive, and easy to build. It can also be applied to the EM material characterization of anisotropic metamaterials.

3.2.4 Asymmetrical Stripline-Based Method

In this method (Gomez et al. 2011), the measurement cell consists of an asymmetrical stripline with a central conductor and two ground planes as shown in Fig. 3.13. In order to concentrate most of the EM energy in the region where the material under test (MUT) is placed, the conductor strip is in the proximity of the inferior ground plane. In this region, the magnetic field is perpendicular to the inclusions whereas electric field is parallel to the inclusions. The amplitude being constant for both fields over the entire sample. The propagation of a quasi-TEM mode can be considered for the analysis. It is noted that the width of the metamaterial sample must not exceed the width of the central conductor to avoid edge effects so as to ensure homogenous field distribution.

VNA is used for the measurement of the S-parameters of the material loaded measurement cell. After the electromagnetic analysis of the cell, the effective parameters of the material can be obtained from the measured S-parameters of the cell. The extraction of material parameters (ε_{eff} and μ_{eff}) is facilitated by two different theoretical approaches using quasi-static approximation.

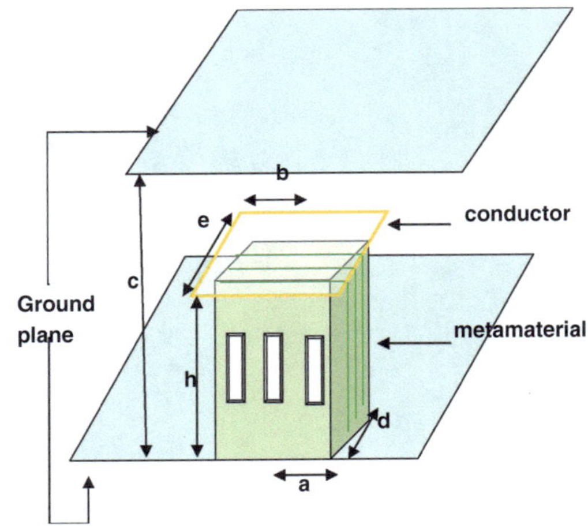

Fig. 3.13 Schematic of asymmetrical measurement cell

(i) **Variational method**: The cross section of the measurement cell is divided into horizontal layers, which represents different media (representing air, material and air gaps) of the structure (Fig. 3.14a). Homogenizing the transversal section of the stripline and representation of the same with an effective permittivity and permeability are shown in Fig. 3.14b. Capacitance and inductance per unit length of the line can be evaluated by Green's function and the theoretical values of effective parameters can be represented with loaded and unloaded C and L. By using Nicholson–Ross technique, extraction of ε_{eff} and μ_{eff} can be done from the scattering parameters of the cell.
Thus, an inverse problem is applied for obtaining intrinsic parameters of the material used in the measurement. The evaluation of ε and μ are obtained by an optimization procedure, which matches the theoretical and practical values. In this method, the length of the sample is considered to be infinite in the

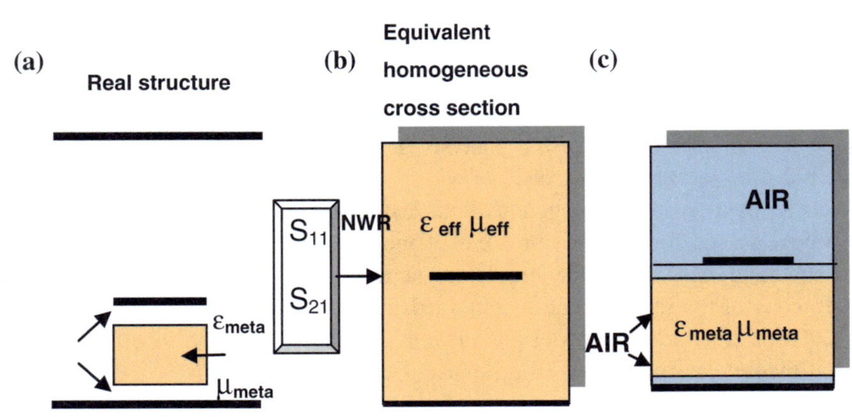

Fig. 3.14 Variational method

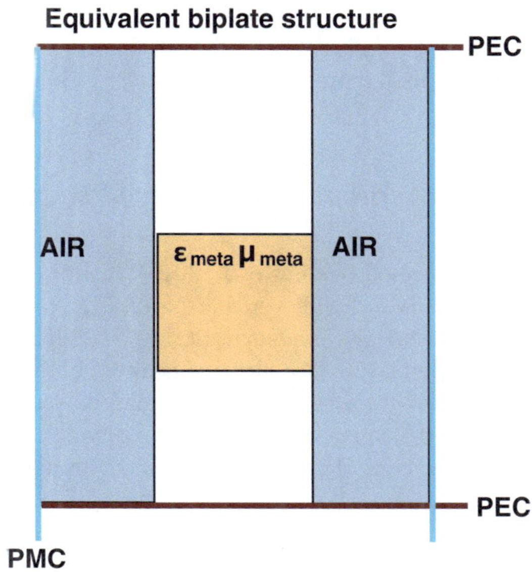

Fig. 3.15 Quasi-static method

transversal direction. This assumption could not be applied as the configuration of the electromagnetic fields outside the coverage of central strip is not considered for measurement purpose. Hence this method requires an additional procedure to obtain the constitutive parameters of the metamaterial.

(ii) **Quasi-static method**: This quasi-static method is based on transmission line theory. The geometry of the cell is such that the concentration of EM energy is below the centre conductor. Hence, the cross section of the cell is represented as a two conductor structure with boundary conditions (shown in Fig. 3.15). It is assumed that only quasi-TEM mode is propagated in the structure. *S*-parameters of the cell can be obtained in terms of capacitance and inductance per unit length of line using transmission line analysis. This inverse problem provides an expression that relates values of ε_{eff} and μ_{eff} to that of scattering parameters. Due to the presence of air gaps (a) between the sample and ground plane and (b) between metamaterial and central conductor in the measurement cell, a correction based on *Effective Medium Approximation* is required.

The final expressions are given below:

$$\mu_{\text{r}} = \frac{\omega\mu_o h_{\text{a}}(1-R) - Z_o W/(\gamma)(1+R)}{\omega h_{\text{s}}\mu_o(R-1)} \tag{3.69}$$

$$\varepsilon_{\text{r}} = \frac{h_s(\gamma^2)}{\omega^2\mu_o\varepsilon_o(h_s\mu_{\text{r}}+h_a) - h_a\gamma^2} \tag{3.70}$$

where ω is the angular frequency, h_a is the height of the air gap, h_s is the height of the sample, Z_o is the characteristic impedance of the cell, R is the reflection coefficient, γ is the propagation constant and W is the width of the strip conductor.

3.2.5 Adjustable Height Stripline Method

In this method (Gomez et al. 2013), the measurement cell consists of a symmetric stripline fixture with variable position ground planes (Fig. 3.16). This particular structure enables in adjustment of the cell height for the measurement of materials of different sizes. Two identical samples of the material under test are placed above and below the central strip. The EM field distribution required for metamaterial characterization is obtained at the central area around the stripline in the measurement cell. Here, the electric field is parallel to the metallic inclusions and

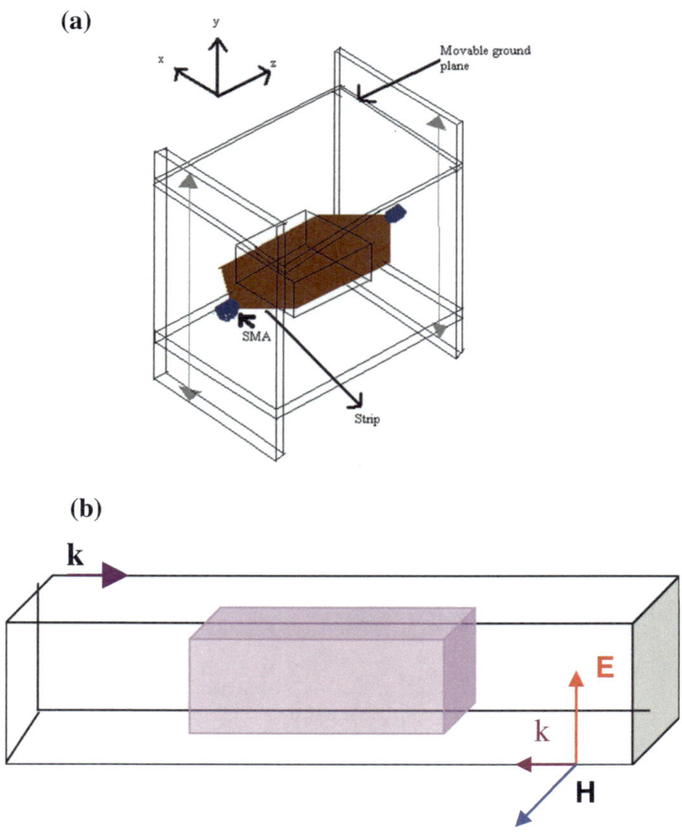

Fig. 3.16 **a** Stripline-based measurement cell, **b** metamaterial unit cell

magnetic field is perpendicular to it. The S-parameters of the sample-loaded measurement cell are measured using Vector Network Analyzer (VNA). Permittivity and permeability of the metamaterial are retrieved from the measured S-parameters based on EM analysis of stripline structure using quasi-static assumption. The length and width of the metallic strip are 50 and 17 mm, respectively. The distance between the strip and both the ground planes vary from 1 to 6 mm. SOLT (Short, Open, Load, Thru) technique is used for calibration of VNA system. An analytical post-procedure is required to account for tapering of stripline, signal attenuation and impedance mismatch of the cell.

The advantage of adjustable stripline fixture is that it allows the measurement of metamaterials composed of several numbers of unit cells in x, y and z directions. Further, it facilitates to evaluate the effect of interactions between successive unit cell structures.

3.3 Free-Space Methods

Non-invasive characterization of materials has drawn much attention for composite structures (like metamaterials) with unusual dispersion properties (Ran et al. 2005; Varadan and Tellakula 2006). Metamaterials are generated from a cyclic repetition of layered structures. Since the modes supported by metamaterial structures are extremely sensitive to external perturbations, non-invasive characterization techniques are preferred for such materials. Hence, free-space measurement (FSM) methods (Musil and Zacek 1986; Ghodgaonkar et al. 1990; Gagnon et al. 2003; Arjavalingam et al. 1990; Carin et al. 1993) are best suited for metamaterials due to their non-invasive nature and wide band characterization.

3.3.1 Synthetic Gaussian Aperture-Based Non-invasive Metamaterial Characterization System

In this technique, a virtual transmitting aperture is synthesized to produce a tight spot Gaussian beam that is focused on the sample under test (Chung et al. 2009). Here, a synthetic aperture concept is employed, which indicates the beam impinging on the sample is synthetically collimated by post-processing a set of individual measurement responses over a desired virtual aperture. The schematic of synthetic FSM (Weir 1974; Baker-Jarvis et al. 1990) system is illustrated in Fig. 3.17.

The transmitter (T_x) probe is moved along the designated scan surface (planar or spherical) to generate the synthetic aperture, while the sample and receiving probe (R_x) remain stationery. For each T_x probe location, the corresponding S-parameters are measured and stored. Using appropriate weighting of each detected signal, the

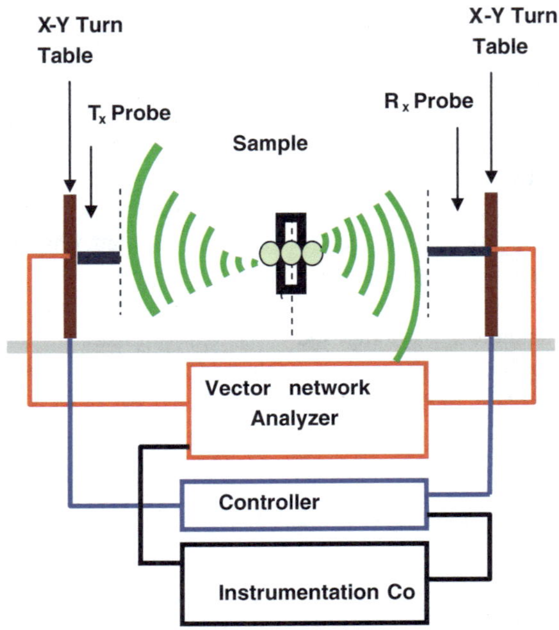

Fig. 3.17 Schematic diagram of synthetic aperture-based FSM

Gaussian beam is synthesized with appropriate weights of each detected signal. This facilitates the generation of a large Gaussian distributed aperture and the difficulties associated with the usage of a large conventional lens are avoided.

Summing the individual S_{21} measurements corresponding to each probe location, the total transmission coefficient is obtained by

$$S_{21}^{\text{total}} = \frac{V_2^{\text{total}}}{V_1^{\text{total}}} \tag{3.71}$$

here, $V_1^{\text{total}} = 1/N \sum_{i=1}^{N} V_1^i$. V_1^i represents the probe excitation voltage at the ith location of T_x probe and V_2^{total} is the normalized voltage measured at the R_x antenna when all the probes are transmitting. Further, V_2^i is the voltage at the receiving end considering the transmission from ith probe only.

Thus,

$$S_{21}^i = \frac{V_2^i}{V_1^i} \tag{3.72}$$

In order to form a Gaussian beam, it is required to weight each S_{21}^i with complex Gaussian coefficients U^i giving

$$V_2^{\text{total}} = \frac{1}{\sum_{i=1}^{N} U^i} \sum_{i=1}^{N} V_2^i U^i = \frac{1}{\sum_{i=1}^{N} U^i} \sum_{i=1}^{N} S_{21}^i V_1^i U^i \tag{3.73}$$

Simplifying the above equation by assuming identical probe voltage at each probe location, one gets

$$V_2^{\text{total}} = \frac{V_1}{\sum_{i=1}^{N} U^i} \sum_{i=1}^{N} S_{21}^i V_1^i U^i \tag{3.74}$$

From Eq. (3.74), the final (plane wave) transmission coefficient

$$S_{21}^{\text{total}} = \frac{1}{\sum_{i=1}^{N} U^i} \sum_{i=1}^{N} S_{21}^i U^i \tag{3.75}$$

Then the measured S_{21}^{total} data with sample under test is normalized by the measured S_{21}^{total} data without sample to eliminate the propagation phase.

This method minimizes diffractions from sample edges facilitating more accurate transmission measurements. Additionally, in contrast to the fixed FSM, it is not necessary to precisely design a lens as the Gaussian weighting process synthetically reproduces a focused beam. This system can easily be integrated to planar and spherical probing systems. It is also suitable for lab surroundings with spurious reflections as the synthesized beam eliminates the scattering from nearby structures.

3.3.2 Miscellaneous Methods

In addition to the conventional measurements based on S-parameters, some miscellaneous methods have been reported elsewhere to determine the complex permittivity and permeability of metamaterial structures.

3.3.3 Waveguide Input Impedance Method

In this method (Hrabar et al. 2006), a single-negative (SNG) material loaded in a waveguide is considered as equivalent evanescent transmission line and the input impedance of this line is assumed to be equal to the wave impedance. This method uses only one reflection measurement for the estimation of complex permittivity or complex permeability of the sample.

The experimental setup consists of SNG metamaterial having either an isotropic scalar permittivity or uniaxial anisotropic permeability filled inside a rectangular waveguide as shown in Fig. 3.18a. This waveguide can be considered as a two-wire

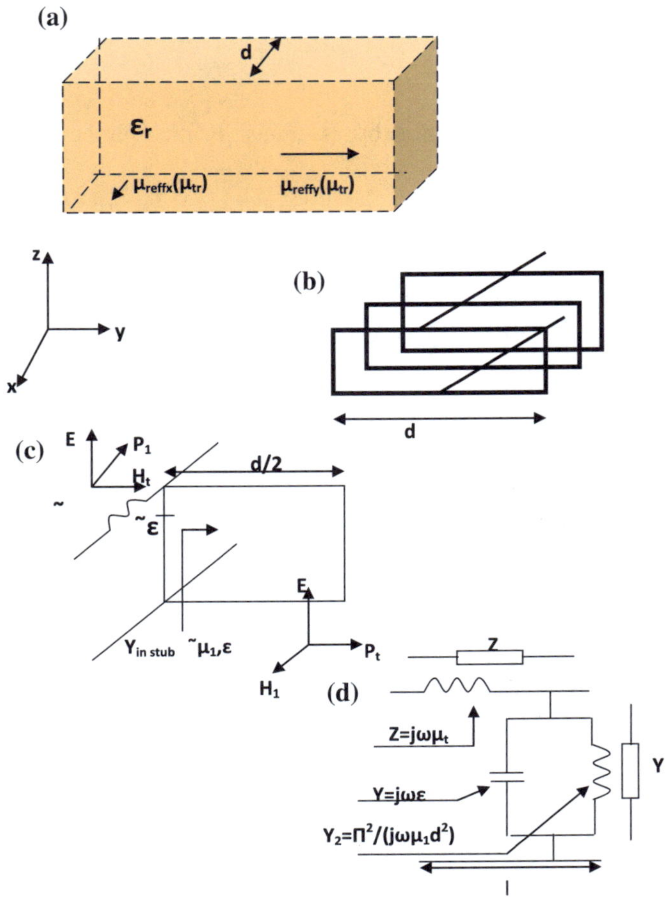

Fig. 3.18 **a** Metamaterial loaded waveguide section, **b** equivalent transmission line representation of a rectangular waveguide, **c** distributed inductance and capacitance of the SNG structure, and **d** equivalent circuit representation of the measurement cell

TEM transmission line loaded with an infinite number of short-circuited stubs (Fig. 3.18b). The scalar permittivity of the SNG metamaterial is represented by the distributed capacitance of the main line.

The energy flow along the waveguide is represented by the longitudinal component of the Poynting vector P_1 as shown in Fig. 3.18c. Due to this, the transverse magnetic field vector H_t exists. This in turn results in the generation of the distributed inductance associated with the transversal permeability (μ_t). Similarly, the energy flow along the stub generates longitudinal magnetic field vector H_1 and hence the distributed series inductance of the stub associated with the longitudinal permeability (μ_1). It is noted that the input admittance of a short-circuited stub can be represented by a parallel tank circuit. Thus, the differential section of the waveguide filled with a general metamaterial can be represented by the equivalent circuit as shown in Fig. 3.18d.

The wave impedance (Z_w) of the waveguide filled with a general lossless metamaterial is given by

$$Z_w = \pm\sqrt{\frac{Z}{Y}} = \pm\sqrt{\frac{\mu_o\mu_{tr}}{\varepsilon_o\varepsilon_r\left[1-(f_c/f_o)^2\right]}} \tag{3.76}$$

$$Y = Y_1 + Y_2, \quad f_c = \frac{f_{co}}{\sqrt{\varepsilon_r\mu_{lr}}}, \quad f_{co} = \frac{mc}{2d}, \quad m = 1, 2, 3, \ldots \tag{3.77}$$

where Z is the impedance per unit length of the waveguide, Y_1 and Y_2 are the shunt admittances, d is the transverse dimension of the waveguide, f is frequency of the signal, d stands for the waveguide width, c is speed of light, f_o and f represents the cut-off frequencies of an empty waveguide and the same waveguide filled with material, respectively.

In case of lossy ENG metamaterial, the unknown complex effective permittivity $\varepsilon_r = \varepsilon_r' - j\varepsilon_r''$ can be extracted directly from Eq. (3.77) under the assumption that

$$\varepsilon_{\text{reff}}' = \frac{1}{\omega\varepsilon_o}\left\{\text{Im}\left[\frac{j\omega\mu_o}{Z_w^2} + \frac{\pi^2}{d^2\omega\mu_o}\right]\right\} \tag{3.78}$$

$$\varepsilon_{\text{reff}}'' = \frac{1}{\omega\varepsilon_o}\text{Re}\left[\frac{j\omega\mu_o}{Z_w^2}\right] \tag{3.79}$$

In a similar way, it is possible to extract the complex permeability for lossy uniaxial MNG metamaterial based on only one measurement. The real and imaginary parts of unknown complex permeability can be written as

$$\mu_{\text{reff}}' = \frac{1}{\omega\varepsilon_o}\text{Im}\left\{\left[\omega\varepsilon_o + j\frac{\pi^2}{d^2\omega\mu_o}\right]Z_w^2\right\} \tag{3.80}$$

$$\mu_{\text{reff}}'' = \frac{1}{\omega\varepsilon_o}\text{Re}\left\{\left[\omega\varepsilon_o - j\frac{\pi^2}{d^2\omega\mu_o}\right]Z_w^2\right\} \tag{3.81}$$

In this way, we can simply measure the wave impedance Z_w, and directly calculate the unambiguous complex permeability μ_r.

3.3.3.1 Q-Based Methods for Resonant Metamaterials

Most of the metamaterials are based on magnetically or electrically self-resonant unit cell (or resonant particle) structures. Using the relationships between resonance parameters, an accurate estimation of the frequency dependent effective material parameters (permittivity/permeability) of resonant metamaterial can be carried out by considering analytical or experimental estimations of the Q-factor of the resonant unit cell (Cummer et al. 2008).

There are two distinct resonant responses in a metamaterial, each one having its own quality factor. These are represented by Q_{part} and $Q_{\text{mat},}$ respectively (Pendry et al. 1999). Q_{part} is for individual resonant cell and Q_{mat} is for magnetic (permeability) or electric (permittivity) material response of bulk metamaterial. The closeness of these two quality factors can be shown through simulations. In metamaterial, individual resonators are aligned in such a way that the coupling is significant but not dominant. Further, all the coupled resonant cells will resonate collectively at a desired frequency, which is close to that of a single resonant cell. During such resonance phenomena, the losses are not shifted from one resonance to another and no additional loss mechanisms are created.

The resonant behaviour of a collection of individual resonators, which have effective magnetic material response, is defined by

$$\mu = 1 + \chi_{\text{m}} = 1 + \frac{F\omega^2}{\omega_o^2 - \omega^2 + j\omega\omega_o/Q_{\text{mat}}} \tag{3.82}$$

where ω_0 is the resonant frequency of the effective magnetic response and μ is the relative permeability of the medium. The loss part is represented by Q_{mat} and $\chi_{\text{m}} = \mu - 1$ is the susceptibility. If the particles are not sufficiently small compared to the wavelength, the functional form of μ will break and spatial dispersion becomes important while loop radiation factor cannot be completely neglected. However, in most practical cases, $Q_{\text{part}} \approx Q_{\text{mat}}$. It is observed that the value of μ also depends on F. For the planar loop resonator configuration, $F = \mu_0 A_{\text{loop}}^2/(V_{\text{cell}}L)$. Here A_{loop} is the area enclosed by single loop resonator, V_{cell} is the volume of a unit cell, and L is the inductance of a single loop resonator. Rearranging of Eq. (3.82) will provide equations for key metamaterial properties in terms of Q and F

$$|\chi_{\text{r}}| = \frac{F\omega^2|\omega_o^2 - \omega^2|}{\left(\omega_o^2 - \omega^2\right)^2 + \omega_o^2\omega^2/Q^2} \tag{3.83}$$

$$|\chi_i| = \frac{F\omega_o\omega^3/Q}{\left(\omega_o^2 - \omega^2\right)^2 + \omega_o^2\omega^2/Q^2} \tag{3.84}$$

$$|\chi_{\text{m}}|^2 = \frac{F^2\omega^4}{\left(\omega_o^2 - \omega^2\right)^2 + \omega_o^2\omega^2/Q^2} \tag{3.85}$$

Using Eqs. (3.84) and (3.85), it can be shown that

$$\frac{|\chi_i|}{|\chi_{\rm m}|^2} = \frac{\omega_0/\omega}{FQ} \tag{3.86}$$

Thus, the susceptibility loss tangent can be represented as

$$\left|\frac{\chi_i}{\chi_{\rm r}}\right| = \frac{(\omega_o/\omega)|\chi_{\rm r}|}{FQ}\frac{|\chi_{\rm m}|^2}{|\chi_{\rm r}|^2} \tag{3.87}$$

It is noted that susceptibility loss tangent is different from permeability loss tangent. Equation (3.87) can be simplified by considering some approximations. The required magnetic material response is attained at frequencies around 5–10% above or below the resonant frequency and $\omega_0/\omega = 1$. Furthermore if losses are not too severe then $|\chi_{\rm m}|^2 \approx |\chi_{\rm r}|^2$. Then the susceptibility loss tangent is approximated as

$$\left|\frac{\chi_i}{\chi_{\rm r}}\right| \approx \frac{|\chi_{\rm r}|}{FQ} \tag{3.88}$$

The neglected factor $\omega_0|\chi_{\rm m}|^2/\omega|\chi_{\rm r}|^2$ (provided $|\chi_i| < |\chi_r|$) is not far from unity and the approximation is accurate, provided that Eq. (3.88) gives the susceptibility loss tangent in terms of three parameters: preferred real part of susceptibility, geometry-dependent factor F, and quality factor Q of individual resonant cell.

Then, the magnetic loss tangent can be written as

$$|\tan\delta_{\rm m}| = \frac{|\chi_i|}{|1+\chi_{\rm r}|} \approx \frac{|\chi_{\rm r}|^2}{{\rm FQ}|1+\chi_{\rm r}|} \tag{3.89}$$

Similarly, the electric permittivity and electric loss tangent can be calculated.

References

Arjavalingam G, Pastol Y, Halbout JM, Kopcsay GV (1990) Broadband microwave measurements with transient radiation from optoelectronically pulsed antennas. IEEE Trans Microw Theory Tech 38(5):615–621

Baker-Jarvis J, Vanzura EJ, Kissick WA (1990) Improved technique for determining complex permittivity with transmission/reflection method. IEEE Trans Microw Theory Tech 38 (8):1096–1103

Baker-Jarvis J, Janezic MD, Grosvenor JH Jr, Geyer RG (1993) Transmission/reflection and short-circuit line method for measuring permittivity and permeability. National Institute of Standards and Technology, Technical Note 1355-R

Boughriet A-H, Legrand C, Chapoton A (1997) Noniterative stable transmission/reflection method for low-loss material complex permittivity determination. IEEE Trans Microw Theory Tech 45:52–57

Boybay MS, Kim S, Ramahi OM (2010) Negative material characterization using microstrip line structure. In: Proceedings IEEE antennas and propagation society international symposium, pp 1–4, 11–17 July 2010
Buell K, Sarabandi K (2002) A method for characterizing complex permittivity and permeability of metamaterials. In: Proceedings of IEEE antennas and propagation society international symposium, Texas, pp 408–411, 16–21 June 2002
Carin L, Agi K, Kralj D, Leung KM, Garetz BA (1993) Characterization of layered dielectrics with short electromagnetic pulses. IEEE J Quantum Electron 29:2141–2144
Chalapat K, Sarvala K, Li J, Paraoanu GS (2009) Wideband reference-plane invariant method for measuring electromagnetic parameters of materials. IEEE Trans Microw Theory Tech 57:2257–2267
Chen H, Zhang J, Bai Y, luol Y, Ran L, Jiang Q (2006) A waveguide-based retrieval method for measuring complex permittivity and permeability tensors of metamaterials. Opt Express 14 (26):12944–12949
Chung JY, Sertel K, Volakis JL (2009) A non-invasive metamaterial characterization system using synthetic gaussian aperture. IEEE Trans Antennas Propag 57(7):2006–2013
Crowgey BR, Tang J, Rothwell EJ, Shanker B, Kempel LC (2015) A waveguide verification standard design procedure for themicrowave characterization of magnetic materials. Prog Electromagn Res 150:29–40
Cummer SA, Popa BI, Hand TH (2008) Q-Based design equations and loss limits for resonant metamaterials and experimental validation. IEEE Trans Antennas Propag 56(1)
Damascos NJ, Mack RB, Maffet AL, Parmon W, Uslenghi PLE (1984) The inverse problem for biaxial materials. IEEE Trans Microw Theory Tech 32(4):400–405
Fenner RA, Rothwell EJ, Frasch LL (2012) A comprehensive analysis of free-space and guided-wave techniques for extracting the permeability and permittivity of materials using reflection-only measurements. Radio Sci 47(1):1004–1016
Gagnon N, Shaker J, Berini P, Roy L, Petosa A (2003) Material characterization using a quasi-optical measurement system. IEEE Trans Instrum Meas 52(2):333–336
Ghodgaonkar DK, Varadan VV, Varadan VK (1990) Free-space measurement of complex permittivity and complex permeability of magnetic materials at microwave frequencies. IEEE Trans Instrum Meas 39(2):387–393
Gomez S, Chevalier A, Queffelec P (2011) Asymmetrical stripline based method for the electromagnetic characterization of metamaterials. In: Proceedings of progress in electromagnetics research symposium, PIERS, Morocco, pp 305–308, 20–23 Mar 2011
Gomez S, Queffelec P, Chevalier A (2013) Metamaterials microwave measurement using an original adjustable height stripline. In: Proceedings of 7th international congress on advanced electromagnetic materials in microwaves and optics (Metamaterials), Bordeaux, France, pp 22–24, 16–21 Sept 2013
Hasar UC (2008) Two novel amplitude-only methods for complex permittivity determination of medium- and low-loss materials. Meas Sci Tech 19:055706
Hasar UC, Westgate CR (2009) A broadband and stable method for unique complex permittivity determination of low-loss materials. IEEE Trans Microw Theory Tech 57:471–477
Hrabar S, Benic L, Bartolic J (2006) Simple experimental determination of complex permittivity or complex permeability of SNG metamaterials. In: Proceedings of the 36th European microwave conference, pp 1395–1398, Sept 2006
Liu X-X, Powell DA, Alu A (2011) Correcting the Fabry-Perot artifacts in metamaterial retrieval procedures. Phys Rev B 84:235106
Lozano-Guerrero AJ, Clemente-Fernandez FJ, Monzo-Cabrera J, Pedreno-Molina JL, Diaz-Morcillo A (2010) Precise evaluation of coaxial to waveguide transitions by means of inverse technique. IEEE Trans Microw Theory Tech 58(1):229–235
Musil J, Zacek F (1986) Microwave measurement of complex permittivity by free space methods and their application. Elsevier, Amsterdam. ISBN-13:978-0444995360

Nesimoglu T, Sabah C (2014) Characterization of metamaterials using a new design and measurement technique for microstrip circuit application. In: 14th Mediterranean proceedings IEEE microwave symposium (MMS), Marrakech, pp 1–4, Dec 2014

Nicolson AM, Ross GF (1970) Measurement of intrinsic properties of materials by time-domain techniques. IEEE Trans Instrum Meas 19:377–382

Pendry JB, Holden AJ, Robbins DJ, Stewart WJ (1999) Magnetism from conductors and enhanced nonlinear phenomena. IEEE Trans Microw Theory Tech 47(11):2075–2084

Qi J, Kattunen H, Wallen H, Sihvola A (2010) Compensation of Fabry-Perot resonances in homogenization of dielectric composites. IEEE Antennas Wirel Propag Lett 9:1057–1060

Ran L, Huangfu J, Chen H, Zhang X, Cheng K, Grzegorezyk TM, Kong JA (2005) Experimental study on left-handed metamaterials. Prog Electromagn Res PIER 51:249–279

Varadan V, Tellakula AR (2006) Effective properties of split-resonator metamaterials using measured scattering parameters: effect of gap orientation. J Appl Phys 100:034910

Weir WB (1974) Automatic measurement of complex dielectric constant and permeability at microwave frequencies. Proc IEEE 62(1):33–36

Williams DF, Wang JCM, Arz U (2003) An optimal vector-network-analyzer calibration method. IEEE Trans Microw Theory Tech 51(12):2391–2401

Yousefi L, Boybay MS, Ramahi OM (2011) Characterization of metamaterials using a stripline fixture. IEEE Trans Antennas Propag 59(4):1245–1253

Chapter 4
Summary

The EM characterization of metamaterials may be performed using (a) analytical methods, (b) field averaging methods, and (c) experimental methods. It is observed that the analytical and field averaging methods based on simulation tools often lead to erroneous results, while retrieving the constitutive parameters of metamaterials. Hence experimental techniques are essentially required for accurate EM characterization of metamaterials as the reliability of hardware realization of metamaterial based sub-systems depend extensively on the intrinsic properties. In this regard, a review of measurement techniques for EM material characterization of metamaterials has been carried out. The salient features of each method have been discussed in detail in this report and an overview of the salient features of the methods considered are given in Table 4.1.

R.U. Nair et al., *EM Material Characterization Techniques for Metamaterials*,
SpringerBriefs in Computational Electromagnetics
https://doi.org/10.1007/978-981-10-6517-0_4

Table 4.1 Overview of EM material characterization techniques

EM material characterization techniques	Methods	Remarks
Waveguide based methods	Waveguide based transmission/reflection measurement technique	The metamaterial sample placed in a waveguide system have dimensions similar to that of the waveguide. The complex reflection and transmission coefficients are measured using vector analyser for extracting the ε and μ
	Waveguide based retrieval method	In this method, measurements are done with different orientations of metamaterial sample in the frequency range corresponding to TE_{10} mode
	Waveguide verification design procedure	Waveguide standard is introduced for the accurate electric and magnetic properties of materials at microwave frequencies. For the characterization of metamaterials these standards act as alternative material. Since the characterised material exhibits low loss properties, it can be used to design a radome
	An estimated method for resolving accuracy problem for ε/μ values near λ/2 resonance in T/R technique	This method helps in the determination of both permittivity and permeability of a low loss dielectric material at around half wave resonance. The material characterised in this method reveals low loss properties, a radome can be designed with the help of this material
Stripline methods	Microstrip line method	The EM characteristics of the metamaterial, placed below the microstrip line is obtained. The characteristic impedance and propagation constant are a function of ε and μ. This method is a wide band characterization technique for meta materials and also is cost effective
	Measurement technique for microstrip circuit application	S-shaped metamaterial with feedback transmission line is constructed so as to maintain a characteristic impedance of 50Ω. The S-parameter of the metamaterial is generated and the material characteristics are obtained
	Stripline fixture method	In this method, different types of metamaterial structures are characterized including SNG and DNG structures. The sample material is used as a substrate to the stripline and complexε and μ of the sample are extracted from the measured S-parameters
	Asymmetrical stripline method	Effective electromagnetic properties of metamaterial are obtained from this method. Here the stripline is asymmetric and is used for broadband characterization

(continued)

Table 4.1 (continued)

EM material characterization techniques	Methods	Remarks
	Adjustable height stripline method	Here an adjustable height stripline transmission line is developed allowing several layer metamaterial characterization. In this method the microwave behaviour of the metamaterial can be studied very closely to that of infinite medium
Free space methods	Synthetic gaussian aperture based non-invasive metamaterial characterization system	This technique is used for RF material and metamaterial characterization where a Gaussian beam is generated with a tight spot, focused on material under test via synthesis using individually measured response
Miscellaneous methods	Waveguide input impedance method	In this method, SNG material is loaded in waveguide and is considered as evanescent transmission line. Only one reflection measurement is required for estimation of material characteristics. The material characterised in this method lossy in nature, hence it can be used in designing of a RAS (Radar Absorbing Structure)
	Q-based methods for resonant metamaterial	The design parameters of the metamaterial are derived in terms of Q factor so as to obtain the material characteristics. It is shown that the Q factor of the overall effective material is equal to Q factor of the individual resonant particles that makes the overall metamaterial

Author Index

R.U. Nair et al., *EM Material Characterization Techniques for Metamaterials*,
SpringerBriefs in Computational Electromagnetics
https://doi.org/10.1007/978-981-10-6517-0

Subject Index

R.U. Nair et al., *EM Material Characterization Techniques for Metamaterials*,
SpringerBriefs in Computational Electromagnetics
https://doi.org/10.1007/978-981-10-6517-0

Made in United States
Troutdale, OR
09/26/2026

54909971R00044